BACK FARM

➤ *Make your home a homestead* ◅

RAISING

 **PIGS**

Breeds • Housing • Management & Care

"EXPERT ADVICE MADE EASY"

Kim Pezza

》 hatherleigh

Hatherleigh Press is committed to preserving and protecting the natural resources of the earth. Environmentally responsible and sustainable practices are embraced within the company's mission statement.

Visit us at www.hatherleighpress.com and register online for free offers, discounts, special events, and more.

Backyard Farming: Raising Pigs
Text copyright © 2016 Hatherleigh Press

Library of Congress Cataloging-in-Publication Data is available upon request.
ISBN: 978-1-57826-621-0

Cover Design and Interior Design by Carolyn Kasper

Printed in the United States
10 9 8 7 6 5 4 3

CONTENTS

INTRODUCTION

Probably one of the most loved yet most misunderstood animals on the farm is the pig. Most people like piglets well enough, who are indeed known for their "cuteness;" however, most think of the adult pig as a dirty animal. If you've picked up this book, odds are you're thinking about adding some pigs to your own homestead or backyard farm. What this means is you are about to discover that this piggy preconception could not be further from the truth.

The pig is easily one of the smartest animals you'll find on the farm, if not *the* smartest. They can also be the sweetest animal on the farm—or, one of the most dangerous animal (but we'll cover that later).

Ask most pig owners who let their pigs actually *be* pigs—meaning they don't keep their animals in containment for their entire lives. Pigs, left to their own devices, will not only develop strong personalities, but can even outsmart their owners. They also make surprisingly great pets—though for those who have kept pigs, myself included, this fact hardly comes as a surprise. Having bred and raised crossbreed farm pigs for many years, as well as having kept a potbellied pig as a house pet and camping buddy for almost 13 years, I can confidently attest that pigs are by far one of the most intelligent animals that I have owned, and though they can be quite frustrating at times (especially if they get out), they are also ridiculously easy to fall in love with.

Pigs can be a fun, occasionally challenging animal to raise. And once you have raised a few, you can bet you'll want to keep on raising them, year to year. Be warned, however: for a real "hands-on" owner, someone who is only raising one or two pigs for meat, pigs are also a very easy animal to become attached to, and make a pet out of—especially if you have purchased it very young. If small-scale, year-to-year raising of one or two animals for meat is your plan, make sure to keep the end result in mind. Otherwise, you'll likely find slaughter time very difficult, emotionally speaking. While you always want to treat your pigs well, at slaughter time you don't want to feel like you are sending the family pet off to the freezer, either! Finding the balance and keeping perspective is one of the most important keys to becoming a successful pig farmer.

So, grab a seat in your comfiest chair, keep your favorite beverage close at hand, and let's learn some of the beginner's basics for raising pigs!

CHAPTER 1

A BRIEF HISTORY OF PIG FARMING

Pigs, including the domestic pig and the Eurasian wild boar, are from the genus **Sus,** the family of Suidae and sub-family Suinae. They are considered **even-toed ungulates**, meaning an animal with an even number of toes (usually two or four, in this case two). And, although even-toed ungulates are usually herbivores (plant eaters), both the domestic pig and the wild boar are **omnivores**, meaning they eat both plants and meat. Highly intelligent, pigs can be found in art and literature, as well as in religious symbolism. In Europe, the boar has been known to represent a standard charge (an emblem emblazoned on a shield) in heraldry. And perhaps the most well-known symbol the pig represents is its place as the twelfth animal in the Chinese zodiac.

Under the Sus genus Suidae is the domestic pig and common Eurasian wild boar. Some will lump the wild boar (non-Eurasian), warthog, peccary and babirusa into the genus, though this is incorrect.

Sus is thought to include 10 living species, along with a number of extinct species. The living species include:

Palawan Bearded Pig

Bornean Bearded pig

Heude's pig

Visayan Warty (critically endangered)

Celebes (or Sulawesi) Warty pig

Mindoro (or Oliver's) Warty pig

Philippine Warty pig

Wild Boar

Domestic Pig

Javan Warty pig

The Origins of the Domesticated Pig

DNA evidence (taken from various teeth and jawbone samples) dating back approximately 40 million years tells us that the pigs native to the Eurasian and African continents originated in the Near East. Found to be very adaptable, the pig was a perfect subject for domestication.

However, the question of when and where the pig was domesticated depends on the source. Some claim that evidence of pig husbandry goes back to 5,000 bce, with evidence of domestication in China going back to approximately 6,000 bce. Others say that the pig was domesticated from the wild boar in approximately 11,000–9,000 bce (despite remains having been found in Cyprus that date back to 11,400 bce). Still others will argue that domestication occurred in approximately 13,000–12,700 bce, at the Tiger Basin in the Near East. As of yet, there is no clear consensus; regardless, it is clear that the domesticated pig has been with us for quite some time.

Pigs with Near Eastern/Asian genes were introduced in Europe and quickly become popular due to their willingness to eat just about anything. Pigs also reproduced well, with multiple piglets per litter, and their meat was easy to preserve. With the availability of dependable, domesticated breeding stock, European farmers were also able to domesticate the European boar, which ironically led to pigs with the Near Eastern gene dying out. By the 1500s, Europeans had already begun breeding pigs for specific traits, related to the meat they produced, creating the foundation for what would later become the "bacon-type" and "lard-type" categories of pigs.

Soon, pigs began making their way across the sea, to the New World. At first, there were only a handful of animals: eight accompanied Christopher Columbus on his famous voyage; Hernán Cortés and Sir Walter Raleigh brought a sizeable number of hogs with them to North and South America; another 13 followed Hernando de Soto to Florida. Within three years, de Soto's small group of swine had swelled to approximately 700 pigs—a testament to the pig's ability to breed prolifically. (This number doesn't even count those animals that were eaten, those who escaped and became feral, or those animals that were given to the area Native Americans!)

As more colonists came and more colonies were established, more pigs were imported from overseas, primarily from Spain, Portugal, and England. (Exactly what types and breeds were brought over remains unknown due to the lack of documentation at the time.) By the end of the 17th century, the typical farm had 4–5 pigs of their own, raised to provide meat, lard/fat, and income (when the need arose). And the trend continued; documentation during the 1800s (more prevalent and reliable than in previous centuries) tells us that most commonly imported breeds at the time were Big China, Berkshire and Irish Grazier (which we'll be discussing in the next chapter).

Early Pig Husbandry

According to the Livestock Conservancy, pigs were historically allowed to freely forage, cleaning up fields after harvests. Later, pigs began to be fed by-products from dairies and bakeries, as well as other waste sources. At this time, pigs commonly consumed spoiled foods and waste; as a result, they were considered valuable for sanitation. But the most attractive feature of raising pigs was their extremely low level of investment; in terms of time, feed and labor, pigs were easier and cheaper to raise than virtually any other livestock.

Traditionally, American colonists would raise their pigs by allowing them freedom to roam, letting them forage freely in the woods. They would then round up many of them for slaughter in early November (in northern and middle colonies), December (in Virginia), and January (in North and South Carolina). Cool weather was necessary for curing, which helped to ensure that the meats didn't spoil.

Traditionally, pigs fell under two types of classifications: lard and bacon. Animals falling under the **lard classification** were thick and compact, with short bodies and stout legs. They were bred for lard (rendered fat) production, as well as for the desired flavor of their meat (due to the considerable fat content). Those classified as lard were fattened quickly on corn. Today, according to the American Livestock Conservancy, the Mulefoot, Choctaw and Guinea Hog (all critically endangered) are the only known lard breeds left in the United States.

Animals falling under the **bacon classification** were long, lean, and muscular. They were fed high protein, low-energy foods such as small grains and legumes, as well as dairy by-products and leftover baked goods. This diet causes the pig to grow slowly, developing more muscle and less fat than their lard counterparts.

Two examples of bacon breeds are the Tamworth (a threatened heritage breed) and the Yorkshire.

Regardless of classification, everything was (and still is) used in the pig—except for the squeal. Besides the meat and fat used, the bones, hide, bristles, and skin are all used in various forms, markets, and/or products.

Pig Husbandry in the United States

By the mid-1800s, the breeding and raising of pigs had become more centralized, focused in the areas in the United States that produced the most excess corn. As production became more centralized, many of the different breeds from other areas began to decline, with some becoming extinct as early as the 1900s.

Around the same time, commercial slaughterhouses emerged to answer the growing need for quick, efficient processing of the market-ready pigs. Pigs were herded to market along trails, in a process similar to a cattle drive. It is also estimated that as many as 70,000 pigs were driven each year, going from production areas (in places like Ohio) to the markets along the East Coast. All of this is surprising, given that the nomadic natives did *not* drive pigs, due to the difficulty in herding/driving the stubborn animals over distances. From my own experiences raising pigs, my feelings are mixed as far as ease of herding. If only a single pig was out, as long as they were not chased they could usually be put back into their pen fairly easily when the time came. However, when trying to move multiple animals, it was a bit more of a challenge, and normally took two or more people to do so. (And even then we would sometimes need to round up a few "breakaways"!) I found that the best way to herd pigs is *not* to chase them, but rather walk with them, subtly guiding them back into their pen.

In 1887, the railroad company Swift and Company introduced the refrigerated railroad car, allowing slaughterhouses to be established right in the middle of production areas, thanks to their ability to ship the pork rather than having to drive live animals to other regions. Terminal markets also developed in Kansas City, Chicago, St. Joseph, and Sioux City, with packing plants opening near the stockyards.

Thanks to these advancements, the lard industry had grown tremendously by the start of WWII. Lard, along with other fats, was used for things like greasing guns and making dynamite. In the United States, national programs were introduced to save unwanted household fats and grease. Once a certain amount was collected, it was taken to collection centers, including local butchers. However, after the war the lard market went into decline. The market for lard pigs collapsed as farmers began to raise a leaner, more muscular animal. Breeds such as the Poland China, Berkshire, Yorkshire, Hampshire and Duroc were favored, while breeds like the Mulefoot and Choctaw went into steep decline.

Today, it is estimated that up to 75 percent of pigs in the United States come from just three breeds: the Duroc, the Yorkshire, and the Hampshire. Not only does this put the commercial pig market at great potential risk, it also poses a problem for the genetic diversity of pigs. Preserving the genetic diversity of pigs, including heritage (old and/or historic) breeds, is now left mostly to the smaller family farms and homesteads—like you!

So, now that you know a little bit more about where these amazing animals come from, let's take a closer look at some of the breeds you might consider including on your backyard farm.

CHAPTER 2

TYPES AND BREEDS

. . . .

There are more than 60 million swine in the United States, with approximately 8 million or more marketed in the United States on a yearly basis. According to the United Nations Food and Agriculture Organization, there are over 939,000,000 worldwide with over a billion killed on an annual basis. Almost 70 percent of the animals in the United States are found in what is known as the **Corn Belt area**, an area of the Midwest which includes Illinois, western Indiana, Missouri, Iowa, eastern Nebraska, and eastern Kansas (with Iowa being the largest producer in the country). A full quarter of the meat consumed in the United States is pork; pork is also the most commonly consumed meat in the world.

For the breeds we'll be discussing in this chapter, our focus will be on some of the domestic varieties found in the United States. Some may be common, others may be near extinction, and some may be almost impossible to get a hold of without lots of networking. However, all of them can be viable candidates for your farm and your needs, depending on what you may be looking for and/or what your ultimate goal(s) may be.

||||||||||||||||||||||||||||||||||||||||||||||||||||||||||||||||||||||||||||||||

Throughout this book, there will be certain terminology used, relating to pigs and their care. The following brief definitions will assist in your understanding:

Pig: Generally a very young animal; may also be known as a piglet.

Boar: An intact male pig.

A breeding boar.

A newborn piglet. Photo by Maarten under the Creative Commons Attribution License 2.0.

Barrow: A castrated male pig.

Gilt: A young female, usually less than a year old and which has yet to farrow.

Sow: A mature female, usually over 10 months of age, and which has farrowed a litter.

Farrow: This term in used as both a verb and a noun. As a verb, it means giving birth. However, as a noun, it refers to a litter of newborns.

Live Weight: The total weight of the living animal (aka "on the hoof").

Hanging Weight: The total weight of the animal after slaughter, but before butchering. Taken after the blood has been drained, the intestines disposed of, and the head and feet removed.

Market Weight: The target weight for when the animal is ready to go to processing.

Shoat: A young pig, one that is somewhere between weaning and 120 pounds live weight. (You probably won't hear this term unless you get well into breeding.)

Weaner: A young pig at the weaning stage.

Feeder: A young pig, usually between 40-70 pounds live weight. They are usually purchased, sold or kept to feed until reaching market weight.

Finisher: An older pig, usually at least 150 pounds live weight, and in the "finishing" stages of market growth.

Market Hog: An animal ready for processing, usually weighing in at an average of 230-270 pounds.

This is only a brief look at some of the breeds that are out there, and is meant to serve primarily as an introduction. Once you have found the breed or breeds that interest you, you can do more in-depth research, including contacting breed associations, talking with breeders, and even taking in area county fairs to learn about some of these animals up close and personal. (Keep in mind that pigs are very social animals, so for the healthiest, happiest animals, you will need to get at least two!)

Pigs, Swine, or Hogs?

Let's get our facts straight. You may hear pigs called a variety of names, chiefly "hogs" or "swine," terms which are used interchangeably. Are these just generic terms, or is there a real difference? **Swine** actually *is* a generic term, one which can be used to refer to any type of pig. However, there is a difference between a pig and a hog—namely, their size. A **pig** is a term for a very young animal, and includes piglets. A **hog**, however, is usually an older animal, weighing in at least 120 pounds **live weight** (the weight of the living animal).

All that being said, we *do* hear these terms used interchangeably, regardless of age, size, and weight. Technically, we are wrong to do so, though the average layperson knows no different. In general conversation, it probably doesn't make a difference. Just know that if you plan to get into the business side of raising swine, you'll need to know the difference between these terms for selling or showing your pigs.

American Landrace

A relatively new breed, the **American Landrace** originated from the Danish Landrace, a breed hailing from Denmark as far back as 1895. In the early 1930s, the United States Department of Agriculture (USDA) made an agreement with the Ministry of Foreign Affairs of Denmark to purchase 24 of the Danish Landrace for some swine research. The only stipulation was that the animals could not be commercially propagated as a pure breed (due to the fact that the Danish Landrace had given Denmark the prestigious position of chief bacon exporting country).

By 1949, however, the USDA had been released from this breeding restriction agreement, and by 1950 the American Landrace Association had been formed, with the American Landrace established as a crossbreed with Norwegian and Swedish Landrace varieties (with a bit of Poland China thrown in for good measure). Today, there are over 700,000 registered offspring from the initial parent stock.

Raised for ham, bacon, sausage, and pork chops, the American Landrace is a white, medium-to-large animal with a long body, fine hair, a long snout, and heavy, droopy ears that slant forward (the top edges run almost parallel to the bridge of the snout, which is straight). The feet of the American Landrace are called **trotters** and are eaten by people as pig's feet or pig's knuckles. They are called trotters due to the fact that, because the four toes on their foot point down, the animal walks on the tips of their toes instead of using their entire foot.

The American Landrace is the fifth most recorded breed in the United States. Registration requirements include:

- Must be white (any other color will eliminate registration).
- Ears must not be upright.

- There may be no less than six functional teats on each side. Inverted teats will also disqualify from registration.
- No large black spots on the skin of the animal; however, a small bit of black pigmentation may be acceptable.

The American Landrace boar runs between 500–700 pounds, while sows range between 450–600 pounds. The sows are good mothers and good milk producers, with the ability to produce large litters (up to 12) of good-sized piglets, with large birth and weaning weights. The American Landrace crosses well with other breeds.

The digestive system of the American Landrace is similar to humans. Like humans, they must chew their food (adults have 44 teeth, while piglets have 28, which they will lose at about 12 months of age). Today, there are over 700,000 registered offspring from the initial parent stock.

Berkshire

The **Berkshire pig** is a heritage breed originating from England, and is the oldest registered breed in the world. The American Berkshire Association was established in 1875 with "Ace of Spades," a boar bred by Queen Victoria and imported to the United States in 1841.

Berkshires are black with white **points** (legs, face, and tail) and pink skin. They have short necks, stocky, block-like legs, and strong feet.

Standards for Berkshire registration include:
- Black and white coloring with erect ears, which must be notched within seven days of birth. (This is not just for

Berkshires; any pig requiring ear notching must have it done within this time period.)
- White on all legs, tail (unless docked) and face. One white leg point may be missing.
- Cannot have either a solid white or solid black face from the ears up. They should not have a solid black nose (rim).
- No solid white on the ears, although some white may be permitted. Some white on the body is also permitted.

Known for their meat quality, Berkshire pork is tender and juicy with a high fat content, which contributes to its well-known flavor. A hardy animal, Berkshires are good for operations where pasturing is the method of choice.

The Berkshire has an excellent disposition, being both curious and friendly. They are a medium-sized pig, running around 600 pounds. They are good mothers, with litter sizes ranging from 6–12 piglets.

As an interesting side note, "Napoleon," the pig which features prominently in George Orwell's classic *Animal Farm*, is a Berkshire.

Mulefoot

A heritage breed listed by the American Livestock Breed Conservancy as critically rare, the origins of the **Mulefoot** are not very clear. While it is possible that their heritage stems from Asia or Europe, it is more likely that the Mulefoot originated from the pigs that the Spanish brought to Florida in the 1500s (possibly from the same ancestral stock as the Choctaw, discussed later in this chapter). Standardized as a breed by 1900, the Mulefoot gets its name from its solid, non-cloven hoof.

The Mulefoot is mostly black (with some white points possible) with a coat of soft hair. The ears are medium in size and are pricked forward, though they do flop. Although uncommon, some have wattles on either side of the neck (little bits of hanging hair or fur covered flesh). There are currently no known purposes for wattles on animals.

Standards for the Mulefoot include:

- Solid, non-cloven hooves (these hooves help to eliminate the threat of hoof rot).
- Solid black; however, some white points may be acceptable.
- Medium ears that flop, but which should not cover the face.

A compact animal, sows can range from 250–400 pounds and can have litters of 5–6 (at times up to 12) piglets. The boars range from 350–600 pounds. Mulefoots are said to have a gentle disposition, and although they can thrive outdoors, they do well in pastured operations. The Mulefoot is both cold and heat tolerant, and as such acclimates well to most climates. Although they may look like feral (wild) hogs, they are not. Valued for their meat and lard, as well as their relative ease when it comes to fattening, the Mulefoot was common in the Corn Belt and along the Mississippi River Valley.

A critical breed unique to the United States, the Mulefoot's numbers have begun to grow, due in large part to the commitment of breeders wanting to save the breed.

Tamworth

Originating in the United Kingdom in the counties of Stafford, Warwick, Leicester, and Northampton, the **Tamworth** takes its name from Tamworth Village in Staffordshire. Also known as the **Sandy Back** or Tam, the Tamworth was imported to the United States in 1882.

Tamworth piglets. Photo by Amanda Slater under the Creative Commons Attribution License 2.0.

The Tamworth is ginger to red-mahogany in color (although early examples of the breed were said to have been orange and purple). It has an elongated head and a long, narrow body. The neck and legs are long, and the pig has deep sides, a narrow back, and an excellent foot structure. Bristles on the body protect the Tamworth from ultraviolet rays; however, when they molt they will seek shade and mud to protect their body from sunburn. They resemble, but should not be confused for, the Duroc (discussed later in this chapter).

Standards for the Tamworth include:

- Golden-red hair that is straight, fine and abundant. The coat should as free from any black hairs as possible.
- A light jowl, with wide space between the ears and a slight dish face.
- Large, erect ears.
- Curly hair, a coarse mane, a turned up snout or dark spots on the coat are all unacceptable.

A medium-sized pig, the Tamworth is intelligent with a good disposition, although some say that they may become territorial towards other livestock in their pasture. Their coat color allows them to adapt well to climates while protecting themselves from sunburn.

Tamworth boars range in size from 550–820 pounds, while sows go from 440–660 pounds. The sows are considered good mothers, with litter sizes averaging 6–10 piglets. It is said that a 100 percent survival rate is not unusual for a Tamworth litter. The piglets may be ready for slaughter in 25–30 weeks. A bacon pig, the Tamworth is good for crossbreeding.

With approximately 1,000 animals in the United States, the Tamworth is listed by the American Livestock Breed Conservancy as threatened; however, due to the fact that they pasture well, the breed is once again gaining popularity. They are hardy and do well in areas with severe winters. An excellent grazer, the Tamworth will also graze the forage that cattle leave behind in the pasture.

Large Black

Originating in Great Britain, the **Large Black** is also known as the Devon Pig or the Cornwall Black. A critically endangered heritage breed, the Large Black is easily identified by its large size, its black coloring and its black skin. First imported into the United States in the 1920s, more were imported once again in 1985 when the Large Black population began to dwindle. By 1960, the breed had become virtually extinct; however, the Large Black Hog Association is working to keep that from happening.

Large Black pigs. Photo by Amanda Slater under the Creative Commons Attribution License 2.0.

Standards for the Large Black include:
- Must be solid black in color.
- Ears should be large, floppy, and cover the eyes.
- Face and snout should be long and straight.
- Bodies should be long and deep.

Docile, hardy, and easily kept in a well-fenced area (partially due to the fact that their large, droopy ears—besides offering protection for their eyes when rooting—also obscure their vision), and are an excellent choice for pasture-based operations. However, due to their rarity the Large Black may be difficult to find.

Boars run about 700–800 pounds, while sows go from 600–700 pounds. With long fertility periods and excellent maternal instincts, sows will average 8–10 piglets in a litter.

Hereford

Originating in the United States, the **Hereford** was created between 1920–1925 from a cross between the Duroc and Poland China, possibly with some Chester White or Hampshire. The Hereford has a very unique color, and resembles Hereford cattle (red and white).

Standards for the Hereford include:
- Two-thirds of the face must be white.
- Two-thirds of the body must be red.
- There should be no white coloring beyond the mid-shoulder area or over the back.
- No belt (a strip of color that encircles the body like a belt).
- Three white legs.

The Hereford is hardy and docile, with droopy ears and is said to be a good choice for the beginner.

A meat/bacon hog, it is listed as being on the American Livestock Breed Conservancy watch list, with an estimated 2,000 breeding animals. Boars can reach 800 pounds while sows can reach 600 pounds. An average litter size is nine piglets, although litters can sometimes go as high as 13–14.

Poland China

Raised for its meat (ham, bacon, sausage, and chops), the **Poland China** was first bred in Miami Valley, Ohio in 1816. Although its breed influences are up for debate, it seems clear that the Poland China stems from a number of breeds, including the Berkshire and Hampshire. It is said to be the oldest American breed.

The standards for the Poland China include:

- Black, with six white points (face, feet, and switch). There may be a splash of white on the body.
- No more than one solid black leg.
- Floppy ears.
- No belt.
- May not have a sandy/red haired pigment.
- Registry does allow **tail docking** (removal of most of the tail while the animal is still small, much like sheep and some breeds of dogs).

In the 1920s and 1930s, efforts were made to introduce the Poland China into China. The venture met with mixed success, as the pigs were not adapted to the climate of China, and the Chinese farmers themselves were more interested in the manure that the animals produced (for use as fertilizer) than their meat capacity.

Durable, strong, and calm, the Poland China has a big frame and a long, lean muscular body with large floppy ears.

Boars may run 550–800 pounds while sows may go to 500–650 pounds. A prolific breed, 16–17 piglets in a single litter is not uncommon. However, the sow doesn't usually do well raising big litters. Due to her size, the sow may sometimes lay on a few of the little ones in the first few days.

Chester White

Originating in Chester County, Pennsylvania, the **Chester White** (also known as the Chester County White) is a Heritage breed that was developed between 1815 and 1818. It was also known as the Chester County White.

Standards for the Chester White include:

- The animal must be completely white with a thick, full coat.
- It must have a dished face, with medium-sized floppy ears.

A versatile and intelligent breed, the Chester White is used in commercial crossbreeding pork operations. They can gain one pound for every three pounds of grain they are fed; however, due to their color, they need shade in the summer. They are also prone to vitamin C deficiency.

A durable and sound animal, the boars will run 550–800 pounds, while sows will range from 500–650 pounds. Sows are prone to big litters, but are said to be excellent mothers. They also make for good pets!

Hampshire

One of the oldest breeds in America, the **Hampshire** is the third or fourth (depending on who you ask) most recorded breed of pig in the United States.

Standards for the Hampshire include:

- Black, with a white belt encircling the body (including both front feet and legs).
- There can be some white on the snout, provided it doesn't break the rim.

- When the mouth is closed, any white under the chin must be smaller than a quarter.
- White coloration is allowed on the rear leg; however, it should not be above the tuber calcis bone.
- Per the Hampshire Swine Registry Board of Directors, as of July 1, 2014, all sires must be DNA tested for color before any litters can be recorded.

A Hampshire pig. Photo by Jim Champion under the Creative Commons Attribution License 2.0.

The Hampshire has erect ears, good muscles, and rapid growth, although its growth rate is not quite as fast as many other cross-breeds. They are good tempered, although some have reported that the boars are aggressive. I had a very large Hampshire cross breeding boar (he ran close to the 900–1,000 pound mark) and he never showed aggression. In fact, when he "retired" he became a pet! The sows are good mothers, with long breeding lives and an average litter size of nine.

Duroc

An old American breed developed in New England around 1800, the **Duroc** is a large-framed, medium-length red animal. Muscular, with droopy ears that cover their eyes, the modern Duroc is a product of breeding and crossbreeding, being bred for color, size, and meat quality since 1860. Although its ancestry is unclear (all we know is that the original Duroc was a *very* large pig) it is regardless considered a heritage breed.

A young Duroc sow.

Standards for registration include:

- No white feet, and no white spots anywhere (except for end of snout).
- No black spots over 2 inches in diameter, and no more than three spots, regardless of size.

- Clear of ridgeling testicle (cases where the boar has only one testicle).
- At least six functional udder sections on each side.

The Duroc is a hardy, fast-growing animal, as long as they have a good, continual, and nutritious diet. They are good in both cold and warm climates. The sows make for good mothers and produce large litters. The breed is also popular for crossbreeding use and improving other breeds. In fact, the Duroc is called a superior genetic source for improved eating qualities of pork by the National Pork Producers Council Terminal Sire Line Evaluation.

Yorkshire

The American **Yorkshire** is the most recorded swine breed in the United States. Called the Large White in England and York-shire just about anywhere else in the world, they originated in Yorkshire, England in 1761, with the breed first being imported to the United States in 1830. However, the Yorkshire did not find popularity in the States until the 1940s, due mainly to their slow growth rates. The breed "improved" quickly, however, due to selective breeding.

The modern American Yorkshire is white with long, upright ears. The animals are muscular with a straight back and lean meat. Found in nearly every state, the highest concentrations are in Illinois, Indiana, Iowa, Ohio, and Nebraska. It is the typical market hog and used in many crosses.

Standards for Yorkshire registration include:

- At least six teats on each side.
- Completely white hair (any color other than white is grounds for no registration).

- The animal must be free from debilitating conditions, including:
  - Blindness
  - Ruptures/Hernias
  - Hermaphroditism
  - Cryptorchidism
  - Single or abnormal testicles
- There must be virtually no black spots on the skin. Although small spots may be acceptable, they are not desirable for registration.

Yorkshire pigs.

The Yorkshire is known as the "mother breed," due to the durable mother lines which have contributed to the animal's longevity and good-sized carcass. Litter size averages around 11 piglets per litter.

It should be noted that data on this breed has been maintained with great diligence, including sow productivity, growth, and back fat formation. As a result, the Yorkshire is the largest source of documented performance records (of any livestock) in the world.

Gloucester Old Spot

Also known as Gloucester, Old Spot, Gloucestershire Old Spots, Old Spots, or Orchard Pig, the **Gloucester Old Spot** is listed as critical by the American Livestock Breed Conservancy. Originating in England, this breed has only been pedigreed since the early 20th century; however, the Old Spot has been shown in paintings dating back centuries.

The Gloucester Old Spot is white with defined black spots (the registry specifies that there must be at least one spot to be acceptable).

Other standards include:
- Large ears that almost cover the face, and which hang towards the snout.
- Straight, strong legs and a long level back.
- Deep sides, with a thick, full belly and flank from ribs to ham.
- 14 well-placed teats, straight and silky hair, and no coarseness or wrinkles on the skin.

A meat, lard, and bacon pig, and excellent for pasture operations, the Gloucester Old Spot is a docile, intelligent, and prolific breed. Calm, good-natured, and even tempered, boars run from 500–600 pounds, and sows range from 450–500 pounds. With wonderful maternal skills, the sows are excellent mothers, with litter sizes between 6–8 (many averaging 10 or more). The boars are seldom dangerous to the piglets and the piglets can be raised well on pasture. However, due to their pink skin, there needs to be shade available to prevent sunburn.

A final interesting note, the Gloucester Old Spot is genetically and characteristically similar to the extinct Cumberland Pig. In the UK, Gloucester Old Spots are being used in attempts to recreate the Cumberland.

Gloucester Old Spot. Photo by Ricardo under the Creative Commons Attribution License 2.0.

Guinea Hog

Originating in the southern United States, the **Guinea Hog** (also known as the Pineywoods Guinea, Guinea Forest Hog, Acorn Eater, and Yard Pig) was once the most numerous pig found on the southern homesteads. Unfortunately, the Guinea Hog became rarer as these homesteads disappeared; they now survive only in isolated areas in the southeast. With fewer than 200 purebreds known to exist, this once plentiful hog is now on the critically endangered list at The American Livestock Conservancy. However, new herds were established in the 1980s

Guinea Hogs. Photo by Fisherman's Daughter under the Creative Commons Attribution License 2.0.

and there are farmers working to bring the breed back from the brink of extinction.

The exact history of the breed is a bit of a mystery, although it is thought that the foundation stock came from West Africa and the Canary Islands, in conjunction with the slave trade in Virginia. They were documented by Thomas Jefferson as early as 1804. The original strain, now extinct, had hints of red and was called Red Guineas. In fact, a Guinea breeder may occasionally find red highlights on piglets and—on rare occasion—a completely red piglet. Guinea Hogs will forage for their food, consuming rodents, small animals, roots, grass, and nuts. In the past, they were also used to clean out harvested gardens. They were sometimes kept in the yard to keep the house area safe from snakes, as they would kill and eat any snakes they found. Most likely, there were many strains of the Guinea Hog; however, most became extinct.

Said to be a good pig for beginners, the Guinea Hog's physical description varies. Usually a small, gentle, black or blueish-black breed, with upright ears, a hairy coat, and a curly tail, the Guinea Hog is unique to the United States, and suitable for smaller properties. Hardy and efficient, the Guinea Hog is excellent for ham and bacon (as it has exceptionally tender meat) as well as lard (having high lard content).

Guinea Hog boars range in size from 250–300 pounds, while the sows weigh in at around 150 pounds. Their litter size averages between 4–8 piglets, but have been known to reach as many as 14. The sows are known to be good, calm mothers who are easy to work with. The boars may breed as early as 6–8 months of age, and the sows at 8 months.

Red Wattle

The **Red Wattle**, so named for their red color and wattles (the fleshy, hair-covered hanging on each side of the neck) originated in the United States. The Red Wattle actually comes in a variety of shades of red; some will also have black patches or specks of black, some may have red and black hair, and some may be almost completely black.

Although their early history is not known, the modern breed descends from animals from east Texas in the late 1960s and early 1970s. When settlers began favoring the higher fat content breeds, which were necessary for soap and lard, the Red Wattles were left to roam the hills of east Texas. They were extensively hunted, and were thought to be extinct, until H.C. Wengler came across a small herd in the forest. He crossbred two Red Wattle sows with Duroc boars, which resulted in the Wengler Red Wattle line. 20

Red Wattle. Photo by Mark Whitby under the Creative Commons Attribution License 2.0.

years later, Robert Prentice came across another small herd, and developed the Timberline strain of Red Wattles. He then crossed his Timberlines with Wengler's line to create the Endow Farm Wattle Hog.

The head and jowl of the Red Wattle is lean, with a slim snout and upright ears whose tips droop. They have a short body and a slightly arched back, with an average weight of 600–800 pounds (although they can go up to 1200 pounds), and are usually around 4 feet high and 8 feet long.

The Red Wattle is a hardy animal, known for its foraging, rapid growth, and lean meat. They have a mild and gentle temperament and are adaptive to climate. They are excellent for pasture operations and small scale farming, and are disease resistant. The sows are excellent mothers and can have 10–15 in a litter, with the excellent milk quantity needed to sustain such a large litter.

In 1999, the American Livestock Breed Conservancy found there were only 42 breeding animals, with 6 breeders. An association was started which has maintained the pedigree book for the breed since 2012. The Red Wattle is also listed by Slow Food USA and Ark of Taste for the quality and taste of its meat.

Ossabaw Island Hog

The **Ossabaw Island Hog** is a critically endangered hog, found on Ossabaw Island off the Georgia coast (near Savannah). They are usually black, but can be black with white spots or light-colored with black spots. The adults are quite hairy, with heavy bristles on their head, neck, and topline. Their snouts are long and slightly dished, and their heads and shoulders are heavy (which seem out of proportion to the rest of their body, but which actually provides them with speed and agility). Boars will run about 300 pounds while the sows will go to about 200 pounds.

Ossabaw Island Hog. Photo by Cliff under the Creative Commons Attribution License 2.0.

The Ossabaw Hogs are descended from the pigs that were brought over in the 16th century by the Spanish explorers, and which were released on the island. DNA analysis finds that the pigs originated in the Canary Islands, which would have been a stop for the Spanish en route to the New World.

The Ossabaw Island Hog is an unusual and important specimen for scientific study, largely due to the fact that the animal has been isolated on an island—meaning that the Ossabaw Island Hog is the closest existing genetic representation of the historic stock brought over by the Spanish. Also, isolated as they have been, the Ossabaw Island Hog provides scientists with the opportunity to study a long-term natural population. The breed is unique in that it has been shaped purely through natural selection, in a climate that has heat, humidity and (on a seasonal basis) scarcity of food. As a result, these animals can store an amazing amount of body fat to survive the lean times.

The Ossabaw Island Hog is an excellent breed for sustainable or pastured pork production. Unfortunately, it has been discovered that those animals who reside on the mainland have lost some of the adaptations that make this breed so unique. Sadly, those unique and special animals left on the island are in danger of eradication: despite Ossabaw Island being their home for over 500 years, they are now considered invasive and destructive.

Choctaw

The **Choctaw** is another descendant from the hogs brought over by the Spanish, dating back to the 16th century. They share two distinctive characteristics with their Spanish ancestors: the wattles on each side of their necks, and fused toes that form a single, mule-like hoof. An American breed labeled as high priority–critical by the American Livestock Breed Conservancy, today's Choctaw has changed little in appearance in over 150 years.

The Choctaw were used by the Native Americans, European settlers, and in the southeastern United States for over 300 years. In fact, they were said to have accompanied the Native Americans on the infamous Trail of Tears. Today, however, there are only a few hundred left, with most of the herd residing on the Choctaw Nation reservation in Oklahoma.

The Choctaw is built for survival, being quick and agile with heavy forequarters. They forage on berries, acorns, invertebrates, vegetables, and roots, and are still allowed to roam free by the Choctaw tribe. They are, however, periodically rounded up, sorted, and earmarked (usually with the help of the Catahoula Leopard dog, another animal of probable Spanish origin).

Originally raised for meat and lard, boars will run 250–300 pounds, with sows ranging between 150–200 pounds. Litter sizes are six or more. They are black, with concessional white markings and ears that may be either erect or with a slight droop. They are smart, hardy, and very self-sufficient; however, it is said that when kept in confinement, the Choctaw can become quite tame.

Piney Rooter

Only found in the United States, the **Piney Rooter** (also known as the Razorback or Woods hog) trace their ancestry back to early domesticated hogs—descended from those free rangers who would run off or left behind (as when some moved, they would just leave the pigs behind) and were never seen again. These animals in turn adapted and turned feral—which of course resulted in feral offspring.

Usually black, brown, or a combination of the two, the Piney Rooter are hardy and have dark skin, although they have a slower weight gain than their domestic counterparts. While far from a popular choice (and not particularly well-suited to a first-time backyard farmer) there are some small farmers who do raise these pigs. However, these *are* wild pigs, and should not be considered for first-time farmers.

CHAPTER 3

HOUSING
· · · · · · · · · · · · · · · · · · · ·

As with any livestock, there are things that you need to take into consideration when selecting housing for your pigs. How many animals will you have? Will they be pastured or penned? What will they be used for? (Breeders will need more room than those being raised strictly for meat.) What is your climate like? Can you give them easy access to food and water in the pen area? (This is a must, as hungry pigs will soon become escapees.) This chapter will seek to cover the basics of providing adequate space and living areas for your pigs, while keeping them warm, safe, and dry.

Planning Your Pigs' Housing

If you are raising only a few animals for meat, you may be tempted to build a small house and a small pen to keep things easier. However, keep in mind that the smaller the space you give your pigs, the more work you will have in the end—the more confined your animals are, the more labor intensive clean-up will be.

Space-wise, the more space you can give your animals, the better. Ideally, a fully mature pig should have **48–50 square feet of shed space**, provided they *also* have an adjacent outdoor area where they can exercise and lounge. If they do not have an outdoor area, you're looking at **around 100 square foot of space** needed per mature pig.

If you *do* have the space for an outdoor area, this space should be at least 200 square feet in size (400 square feet if your pigs are breeders). Think of this space as giving your animals something like an exercise yard.

How is the **drainage** on the property? If you have multiple spaces to select from, you will want to pick the site with the best drainage. We will discuss this further on in this chapter; however, the better the drainage in the pen/pasture area, the more animals you can accommodate.

If you are pasturing, some say you should be able to run up to 10 animals per acre of good pasture, provided you are practicing **rotational grazing** (moving animals into another, unused part of the pasture when it has been eaten down, so that the area they were feeding in has a chance to recover and regrow before being put back into use), and giving supplemental foods. In other words, pigs do not survive on pasture/foraging alone (see Chapter 4 for more details on feeding your pigs).

Others will tell you to look at your **soil type**, and allow 100 square feet minimum per animal (if your soil is sandy and well drained), or 200–250 square feet per animal if the soil is clay (which is not a good drainage soil). If you have the space, some even recommend as much as 800 square feet per animal! Again, the more you can give them, the happier everyone will be.

However, these numbers should not be looked at as set standards; there are other things to look at, such as drainage or how good the pasture area is. Not to mention that all of this may change from year to year. A year that is too wet or too dry could affect how you will pasture that year and how many animals you will be able to handle per pasture. Basically, while you can still plan this out as carefully as possible, if Mother Nature has other plans, you will need to adapt!

One thing to keep in mind when deciding on location and how many animals to put into one area is that for large animals (like the pig) that have very small feet, their feet have to hold a lot of weight over a small area. As a result, if there are too many animals in too small of a space or in an area of bad drainage, you can wind up with one gigantic mud hole. And while your pigs may like the occasional mud roll to cool off and protect them from heat and sun, it is not healthy for them if their entire pen is muddy.

Photo by Orange Aurochs under the Creative Commons Attribution License 2.0.

Sheds

Although your pigs may prefer to sleep outdoors in warm weather, you will still need some sort of **shed** for them. But while you might think that you'll need a huge building for such big animals, the truth is that you really don't. Pigs tend to sleep quite close to each other, so they don't take up as much space as you might suppose. The biggest thing to pay attention to is that your shed doesn't get a wet, muddy floor (as this can chill your animal), and that you have good ventilation, a good roof, good drainage, and good bedding.

It also helps if the shed has a **floor**. This will help keep your animals dry, even if you end up having some problems with drainage. And, as pigs do not like wet beds, a floor will make them very happy while helping to keep them warm. The floor should be non-slip, which is easy to accomplish—simply attach batons across the floor to give the pigs grip.

When contemplating where to put the shed for your pigs, remember that strategic placement of the shed (along with strategic plantings when and if necessary) will help in two important ways: **temperature control** and **climate control**. Select a location that will give the maximum protection from heat, snow, and winds (a  sheltered, shady area is good). Make sure that your pigs have shade that they can go to outside of the shed so that they can cool off.

Remember, if you are in an area where there are winters, any shade trees will lose their leaves, allowing the sun to reach the sheds during that time of the year. In addition, look for a spot where you can take advantage of cool winds in the summer, but which can be easily protected in the winter. I used to add a heavy flap to the door in the winter, and found that to work quite well.

Another consideration when positioning the shed and yard is to try to place it far enough away from your house to avoid any odors, both from the pen and from wherever you decide to dispose of the dirty bedding (which is excellent for compost!). Another very important issue is **drainage**. Do not allow drainage from the pens to run into rivers, ponds, streams, or around your well, as it can lead to contamination.

Shed Bedding

It is also very important that you keep your shed clean as well as dry, as this will affect not only your pig's health, but its performance as well. For **bedding**, some will say to use **hay** (as opposed to straw), as using straw bedding risks eye infections (if there are seed hulls in the straw). Not to mention that, should the pigs want to snack, they can eat the hay you lay down for them. That being said, I used **straw** for bedding for years, not only with my pigs (and their litters) but with my goats as well, and never ran into any problems. I found that the straw kept the animals warmer and drier; however, this decision is totally up to you, as either will work. And don't worry about the pigs defecating in their shed: pigs will usually not go to the bathroom in their house or bedding.

When I put straw in, I put *lots* of it in (especially in the winter). My pigs loved to burrow right down into the straw during the cold weather. Back when my breeders were smaller, sometimes I wouldn't even be able to see them—they'd burrowed in so deep. But it kept them warm, even in below freezing and sub-zero temperatures. Not to mention it was much safer than using heat or brooding lamps. (My shed also had a low ceiling—although not so low that the pigs couldn't stand comfortably—which I'm sure helped to keep the heat in.)

Ventilation

The shed should have good **ventilation**; at the same time, you'll need to watch out for drafts. Some will **insulate** their roof and walls to prevent drafts; I just made sure that there were no spots in the walls where drafts would be a problem. Look for any cracks or holes in the walls, any open seams…anything that could allow cold air drafts to seep in. Bales of straw piles around the outside of the building, especially at a wall that gets hit hard with wind, will help as well. If you are building the shed yourself, leaving openings under the eaves of the shed roof will also help with ventilation.

Remember that pigs *can* tolerate low temperatures, so long as they are not exposed to drafts. That said, this can also depend on the age of your animals. Very young and very old pigs (although old pigs tend to be pets, and are likely kept indoors) will need extra heat. If you see your pigs, regardless of age, huddling in a pile, or if your animals are shivering or eating more than usual, then they are too cold and need more bedding, and possibly even a **brooding lamp** (if the weather is really bad). Make sure that the brooding lamp is kept a safe distance away from the bedding, be it hay or straw, and the curiosity of the pigs so that a fire doesn't start.

On the flip side, if your pigs are avoiding bodily contact with each other, eating less, are going to the bathroom in areas that are usually clean or you see them panting, then they are too warm and you need to find them an area where they can cool down.

Types of Housing

When deciding on **housing**, there are many options available for your pigs. Some options may be better than others: for example, a **wooden shed** would be better than a metal shed, as without insulation, a metal shed can be cold in the winter and hot in the summer. Even certain plastic ones will work; when it comes right down to it, the choice is yours. Look at your budget, your available space and your size needs. Study what is out there and purchase what is best for you. If you can't find what you want, pig sheds are easy to build, so keep that option in mind, as well.

If you already have a large barn available and are thinking about keeping your pigs contained there, you might want to think about that a bit more. Although keeping your animals contained sounds easier for you and safer for the animals, that's not always the case. Keeping your pigs in the barn all the time is *very* labor-intensive, and you will have a lot of cleaning to do on a daily basis.

While having a pig yard will not eliminate cleaning, a nice big yard means you will have to clean less. And, if you pasture, the manure will simply eventually work its way into the ground. Your pigs also need fresh air to keep them healthy, and keeping them in a closed building will not provide them with the fresh air and ventilation that they want and need.

Although there are many so-called **pig houses** on the market, feel free to look at other alternatives. Some owners will purchase **commercially-made wooden sheds/mini-barns** and make modifications themselves, such as removing the lower panel of one door so that the pigs can go in and out easily without the need to keep those big doors open. Others will replace a bit of the roof with a clear panel, to allow the sun to beat into the shed during the winter. (This could be quite warm in the summer, however, so you may want to cover it in the winter if shade trees won't suffice.)

A mini-barn can be a great option for housing your pigs. Photo by GemTec1 under the Creative Commons Attribution License 2.0.

Most mini-barns and sheds do have floors; however, if they don't, a floor is easy to install.

Another option is a porta-hut. **Metal porta-huts** are metal buildings with a rounded top, and while they *are* used for pigs, if not insulated they can be hot in the summer and cold in the winter. Therefore, porta-huts need to be in spots that stay dry; in fact, they do best on a platform (your pigs will appreciate their "porch" as well!) If you are in an area with snow, however, their rounded roof will help shed snow.

Calf hutches—those large, plastic-like sheds made for calves—may be used as well. With lots of straw or hay, and a heavy covering over the door, these little sheds can stay pretty warm. However, summer heat can play havoc with their internal temperatures, as the built-in vents are not always all that good.

As a result, calf hutches work best in shaded areas during the summer. These sheds are durable and come in a variety of sizes; however, you may need to add a floor.

If you have an old **greenhouse** on the property that you no longer use, you may want to try to incorporate it into your pig's set-up. Although they would be too hot to use for summer housing, they can be a comfy, cozy resting spot on a cold, sunny winter's day.

Building Your Own Pig Shed

Building your own shed is another option—one that is easier than you think. For the years that I had my pigs, they lived in a shed that I built, with an associated yard. I knew that it had to be big enough for my two adult breeders and one litter per year (which averaged about 10 piglets per litter). It was basically three two-thirds sided, with the fourth wall having the door's opening. In the winter, I would put a heavy flap over the door, such as a heavy tarp or rug (in the summer it was left open). I also had a window in the back which again was open in the summer and covered in the winter using a heavy, clear plastic. This allowed daylight to come into the shed, but was positioned so that in the summer, the pigs were not bombarded with the hot sun. The shed was about 20 feet in length and 10 feet front to back. The front of the shed was about 5 feet high and about 4 feet in the back with a metal roof. This slanted roof helped to shed snow and made snow removal easier. It also kept the rain from running into their doorway. Although I could not stand in the building, my adult pigs were comfortable and cleaning was still easy.

Your pigs' shed can be almost any shape or size, as long as you remember to give them enough space, prevent drafts, and allow for good placement and ventilation. They can be ready-made purchased or homemade, as fancy or as simple as you want (although, given how destructive pigs can be, you will probably want to keep it simple). Have some fun designing it, but keep it simple, practical and most important, safe for your pigs. (Some links for shed plans can be found in the Resources section.)

Fencing

Without proper **fencing**, you will be spending a lot of time rounding your pigs up from their adventures in the outside world. Don't let those bodies fool you—pigs are faster than you think and those little ones can squeeze through some pretty tight spots! Meanwhile, the older, larger pigs will work those same spots until they can get themselves through, one way or another!

When trying to decide on the proper fence, the **strength** and **height** of it will usually depend on what type or breed of pig you are keeping. All pigs, once they are grown, are strong, so your fence will need to be able to deal with the physical strength of your animals. If your fence is too weak, a pig can knock it right over simply by leaning on it. Too loose, and they will figure out how to get under it. If the wires are too thin, a large pig could pull on it until it breaks, or else just rip it apart and go through it. Too low, and a pig will jump the fence. I actually watched my full-grown boar jump and clear a temporary fence. He was huge, so I wasn't expecting it. From then on, fences doubled in height at my place.

So, how do you fence in an animal that is strong, fast, and can be difficult to contain? The answer is, you train them to a fence, starting as young as possible. But before that, you must decide what kind of fence you want.

The easiest way to fence in pigs is through the use of **hog panels**. Sold in 16 foot lengths, and anywhere from 34–50 inches high, these are very rigid, metal fence panels which can support the weight of a pig without buckling. The fence panel has larger space openings in the top section, with smaller openings at the bottom, to keep the piglets from going through. The panels may be attached to either fence posts or metal t-posts to form the yard. As long as there is good drainage, these panels will work well. The one drawback of the hog panel is cost; if you are fencing in a large area, the panels can become expensive, in which case you may want to look at wire or field fencing.

Wire or **field fencing** comes in a variety of heights, gauges, and opening sizes. For use with feeder pigs, 32 or 39 should be sufficient. Although, from my own experience I would recommend the higher numbers, especially if you are keeping breeders (or even farm pigs as pets). The gauge of the wire should be thick enough to withstand the punishment your

Photo by Bruce Guenter under the Creative Commons Attribution License 2.0.

pigs will throw at it; 2.5–3 mm should be sufficient. And, though it should go without saying, there is no way that chicken wire of any kind will work!

The fencing may be put up using wooden or metal t-posts. I found that I preferred wooden fence posts, and I reinforced in the post hole with cement. Although I put cement in every hole, many will only reinforce the corner posts with cement, as they are the most important to the integrity of the fence. The wooden fence posts should be 3–5 inches in diameter, and approximately 5–6 yards apart. The closer the posts are, the stronger your fence will be, and the tighter the bottom will be, making it more difficult for the pigs to get under. If you are fencing through any areas where there are dips in the landscape, put the posts closer together in order to make sure that the fence is tight against the ground. When attaching the fence to the posts, pull the fencing as tight as you can.

Some will also run **boards** along the bottom of the fence to keep the pigs from digging out with their snouts. (A pig's snout is very strong and, unlike a dog, which will dig out from under a fence with their feet, a pig will use their snout to root under the fence, and then continue to push and root until they escape!) For my pigs' yard, I found success by sinking the fence 6 inches underground, then sinking a 6 inch **dig guard** (a thin piece of metal that comes in various lengths that you hammer into the ground to prevent the animal from being able to dig out) around the entire bottom inside perimeter of the fence. It gave me double protection, and worked well for me, even with full grown, 4 year old adults (a sow and intact boar). No one ever dug out of that yard!

Pigs have very strong snouts and can use them to root under fences.

If it is in your budget, purchase your fence at least 6 inches higher than you need, and purchase the dig guards. If you can't find the guards, you can easily have them made out of sheets of aluminum.

You may also choose to run a string of **hot wire** (electric fence wire) along the bottom of the fence to keep the curious little piggies from trying to burrow under the fence.

If you are doing only a small area, such as a pig yard, a **wooden**

fence can work as well. If properly done, a wooden fence can be very attractive and give your farm that old-fashioned look. But make sure that the wood is strong enough to stand up to the abuse your pigs may give it. Remember, they like to chew, lean, and dig. Many people will reinforce a wooden fence with either an electric fence, placing hot wires between each rail and under the bottom rail between it and the ground, or lining the interior of the wooden fence with the wire fence. While this does sound like twice the work, in a smaller space it can give extra security as a double fence, especially for someone who is really set on having a wooden fence and don't want to deal with reinforcement via an electric fence.

Chain link fence will *not* make good fencing for your pigs. Chain link is really quite flexible and, because of the way that they are constructed, they cannot be pulled tight enough to keep the pigs from pushing out of the bottom. Although they can be sunk in, this is usually not a feasible route to take. For very young pigs, however, chain link could be viable as a temporary enclosure while building their actual pen, if need be (although ideally the pen should be built before the animals come home). Again, running a hot wire along the bottom will help keep the pigs from pushing under the fence.

Some keepers that I have spoken with have had good luck using just an **electric fence**. Some were using only a couple strands, while others used a portable electric fence so that they could move their animals around (this also helped with pasture management). Although this *can* work, electric fences are usually used for reinforcement of other fencing and as pasture management.

The biggest problem with using only an electric fence is that, while the pigs will get shocked and scared the first time, whether they decide to back off or plow ahead forward and through is up to chance. It is for this reason that some pig farmers will first use fence training.

Fence Training

In **fence training**, you work to teach your pigs to respect the boundaries of the fence, so that even when not electrified, they will not want to charge through. Fence training is exclusively for those using planning to use an electric fence. Using an electric fence hot wire at snout height temporally inside of your regular fence can help, as when the pigs touch the fence, they will get a shock. Nothing harmful, but they will eventually learn. The good thing about fence training with an electric fence inside of your regular yard fence is that, if the shock scares them and they still choose to go through the electric fence, they still will not get loose—they will still be in their main enclosure. If they are in an electric fence only enclosure, you risk an escapee. When you feel that your pigs are trained to your satisfaction, remove the training fence.

At this point, it should be noted that when putting up a few hot wires or a full electric fence, you cannot just attach the wire to the fence post, no matter what type of fencing you use. In order for the fence to work properly, and carry a good charge, you must use insulators on the poles so that the wire is not touching the pole. **Insulators** are just little gadgets made of various materials in various types and colors that attach to the posts. The wires then attach to them, allowing the fencing to work properly.

Unfortunately, the work doesn't end when you are finished putting up the fence. It is very important to keep tabs on your fence. In a pasture situation, walk the perimeter regularly. Check for holes, weak areas in the fence—places that the pigs can get out from, chew through, or get under. Also, check for problems with your electric fence, if you're using electricity. Make sure the connections are not broken, and that nothing came down on it to break the connection. Fix any problems immediately to prevent problems or escapes.

Once you get a good fence up for your pigs (whether for a yard or pasture) and a good pig shed where your pigs can chill out and sleep, not only will your pigs thank you, you'll feel much better. Although a fence doesn't guarantee that one of your little rascals won't find their way out, it *will* make care easier for you, and be much healthier for your pigs.

CHAPTER 4

FEEDING

· · · · · · · · · · · · · · · ·

The truth is, pigs will eat almost anything. Once called "mortgage lifters" on dairy farms, pigs were known to consume any edible garbage they got their hooves on, including excess whey and milk. This omnivorous behavior, along with pasturing, are what made pigs so inexpensive to feed, while still serving as a huge financial asset for the farms themselves. (Not to mention, the pigs could be sold when money was in need!)

So, if you just happen to be a cheese maker, don't be afraid to feed your pigs the leftover whey from your endeavors! You might also find local stores who are willing to save their vegetable scraps or out-of-date (but not moldy or rotten) product for you.

That being said, although pigs will probably eat much of what you give them, that doesn't mean that they *should* eat it. It is actually very easy to feed your pigs incorrectly. As we discussed earlier, pigs *are* omnivorous, which presents the pig owner with a wide range of options for feeding. However, keep in mind that they also need a balanced diet; otherwise, their growth rate can be affected. And, although pasturing is very important to a pig's diet, it cannot be their entire diet, like it can be for cattle.

Creating a Balanced Diet

Pigs require a protein-rich diet, although their diets will vary during their various stages of life. When they are young, pigs will need up to 18 percent protein in their diet, which drops down to 14 percent as they mature and finish. Although not always financially feasible for larger pig herds, smaller farms with only a few animals have  the option of feeding their animals commercial rations. It is also possible to mix feed on the farm. However, no matter what direction is taken, the following ingredients, in whole or in part, are necessary (and found in commercially produced foods):

Grains and By-Products: This is a major source of carbohydrates. For most feeds, this is corn based; however, there may also be sorghum grain (milo), barley, wheat, rye, and oats.

Animal and Plant Protein: Includes protein of animal origin and soybean meal.

Whole Soybeans: Unprocessed and used successfully as part of gestation and lactation diets.

Dried Skim Milk and Dried Whey: Typically included as 10–30 percent of a piglet's diet, at about 10–30 days past their weaning.

Fats and Oils: This includes (animal) grease and tallow, corn and soy oil (vegetable), as well as fat/oil blends.

Vitamins and Minerals: The precise balance of vitamins and minerals necessary depend on the mix, the age group, medical needs, and breeds being considered. Some of these needs may be fulfilled through their feed, but you may find yourself needing to supplement.

Food Additives: Although food additives are non-nutritional, they may be added to improve feed utilization, health, and/or metabolism.

There are many "recipes" online for homemade pig foods, but if you *are* going to make your own food at home, I would suggest a quick consult with your vet to make sure that you are indeed creating a balanced diet for your animals—one that will meet their nutritional needs for the phase of life they are in. For commercial foods, talk with your local feed store. They can assist you in deciding which feeds are best for your current needs. If you are fortunate enough to have a store that will custom mix feeds for you, take advantage of your good fortune and map out a good feed strategy.

As far as feeding pigs meat, there are two schools of thought. Some owners will do so, including innards from other butcherings (deer and chickens; usually not pork), table scraps, etc. Others choose not to feed their pigs meat, as they fear there could be problems if their pigs get a taste for meat. In my experience, having kept my animals on a good, non-medicated, carefully selected commercial feed (supplemented with fruit and vegetables, both whole and scraps), old (but never moldy) breads, hay, and even grass clippings after I mowed, I did not see any particular need to feed my animals meat. As a treat, I might give my pigs a piece of bologna once a month, or perhaps a cookie. Even my breeding pair usually got a small dog biscuit every morning (again, as a treat). But I never fed them "slop" of any description, and never fed them spoiled or moldy foods.

Feeding Your Pigs

When it comes to actually feeding your pigs, most find that using either a trough or heavy rubber bowls work best. Personally, I chose to use bowls for both food and water which worked out very well. They were basically indestructible, no matter what my animals did to them! And in the winter, it was much easier to get the ice out of the water bowl. (I also like to hide treats around the pen for the pigs to look for; it's fun for them, and keeps them stimulated!)

As an important aside, keep in mind that, if you are raising your pigs for meat, in the end, you'll be eating whatever your pigs eat. So, the better you feed them, the better they'll feed you.

When deciding on how much to feed your pigs, the best measurement is **whatever they will eat in 20 minutes, repeated 2–3 times per day.** The feed amount will also depend on your location, the general climate (your animals will eat more in the winter, using some of the energy from their food to keeping themselves warm), and the animals' stage of life. Keep in mind that pigs below their lower critical temperature will require more feed to maintain their body temperature. So, for every 20 degrees below 60°F, increase their feed by at least 1 pound per animal.

There are commercial feeds for each stage of your animals' life. These include:

Starter: This feed contains 20 percent protein, and can be found medicated. Starter feed is for piglets aged 7–10 days old, up through weaning. Starter feed is intended to optimize the piglet's growth performance in the first few weeks after weaning. Weaning is a very stressful time for the babies, and the diet change will add to it.

Some will choose to wean their piglets young, at 10–21 days. This can cause problems, resulting in a decrease in gain, as well as low feed intake, leading to sickness and death. Piglets weaned at this young age should be fed limited soybean meal, due to the fact that they will develop an allergy to the soybean proteins, increasing the chance of diarrhea and further reducing growth rate. However, within two weeks

FEEDING

their tolerance should build up and their performance should improve.

For those weaned at younger than 4 weeks, a complex starter diet has been shown to result in a significant improvement over those fed a simple soy meal starter. A complex starter will contain high levels of dried milk product, specially processed soy product, animal by-product and a highly digestible carbohydrate source. Some studies have also shown that complex starter diets with a high amount of milk product also resulted in improved performance during the grower/finisher phase.

Grower/Finisher: A 15–16 percent protein feed, grower is fed to growing/finishing pigs beginning at 3 weeks after weaning (around 50 pounds) until they reach their market weight. At this stage, the pigs do not require a complex diet, nor do they have the same needs as starter/nursing pigs, but they do require a higher level of amino acids.

Barrows (castrated males) will eat more than gilts (young females, not yet bred). The gilts will have less fat with more muscle, a higher carcass yield, and a better feed conversion than the barrow.

Boars will gain faster and more efficiently, with less back fat than both gilts and barrows. Growing boars should be fed as normal up to 240 pounds, before backing off a bit.

Sow/Lactating: A 16 percent protein feed, to be fed at late gestation and to lactating sows.

The condition of the animal's body will dictate how to feed during gestation. Take a visual appraisal of the sow's body condition, and reassess diet and feeding schedule accordingly:

- Emaciated: Hips and backbones prominent.
- Thin: Hips and backbone easily felt without applying pressure.
- Ideal: Hips and backbone felt only with firm pressure.

✦ 57 ✦

- Fat Hips: Check to see if you can feel the backbone.
- Over-Fat: Hips and backbone heavily covered with fat.

Obviously, a poor body condition will require more feed, especially in lower temperatures, as it becomes necessary to maintain body temperature.

For breeding sows, poor body condition can increase cull rates (discussed in Chapter 7), increase the number of gilts in the sow herd, and decrease the number of sows per year.

For overly fat sows, there is the risk of increased embryonic mortality, increased risk of difficulty when farrowing, a higher piglet mortality (from the sow accidentally crushing a little one), a decreased intake of feed during lactation, lower milk production than normal, and an increased susceptibility to sunstroke.

If the sow is too thin, there can be failures in maintaining estrus, lower conception rates, smaller litter sizes, and downer sow syndrome (spinal injury and bone breakage due to the excess mobilization of minerals from the bones).

Pig feed: A 14 percent adult maintenance feed, pig feed is also used for dry sows, sows in early to mid-gestation, and boars. This is what I kept my pigs on at all times. I used a mix that was free of animal protein, and I also gave my pigs fruits, vegetables, grass, and hay. I had very good luck using this, with my animals keeping good weight and good health.

During gestation, there are a number of factors that will dictate a sow's nutrient requirements. This includes **feed intake** (the nutrients actually consumed daily) and **production level** (for example, a sow with a large litter will require more nutrients than a sow with a small litter).

A common, general-purpose gestation diet would be an average of 4–5 pounds of feed per day, but this will vary with each individual animal. Keep in mind that too little food intake may cause a vitamin and mineral deficiency.

With lactating sows, you need to optimize their milk production while keeping a good nutrient balance. If the balance is off, excessive weight loss and decreased milk production may be a result. For 15–25 pounds of milk per day, the nutrient requirements will be triple that of a sow during gestation. This nutrient intake is directly related to milk production and the nursing piglet's growth rate. A negative nutrient balance may be minimized by either increasing a sow's food intake or nutrient concentration.

Feeding Boars

When providing nutrition to your boar, you are looking to control their weight and body condition while still keeping optimal breeding performance. It is commonplace to feed your boar the same diet as a gestating sow; however, this may affect their libido and semen productivity. A better alternative is a limited sow lactation diet; protein and lysine is higher, while the energy density is similar. Your animal should average 5–6.5 pounds of food per day, again depending on size, temperature, how heavily you use your boar, etc.

Note that not everyone will keep a boar as I did. I've added this side note regarding their nutritional needs, should you be considering keeping a boar.

Water and Hydration

While we have been discussing feed for your animals, we have not yet discussed the most important nutrient in your pig's diet: water! Water comprises 80 percent of a piglet's body at birth, and 50 percent of a market hog's body. Pigs should have free access to water, even before weaning. (Note that you will have to provide your pigs with water dishes that are low enough so piglets cannot drown if one falls in.)

The amount of water needed will vary with age/stage of life, lactation, and whether or not the animal has a high urine output or diarrhea at the time. Note that lactating sows will consume more water, due to their milk's high water content. If you restrict their access to water, it will negatively affect their milk production. In fact, if you restrict water to *any* pig, it will reduce performance and the results could be deadly. Housed in thermo-neutral conditions, a pig will consume 2–3 pounds of water for every pound of dry food consumed. Heat stress or lactation may increase the intake by 4–5 pounds. To give an idea of the amount of water each animal will consume per day:

- Nursery Pig: 1 gallon
- Grower: 3 gallons
- Finishing: 4 gallons
- Gestating: 6 gallons
- Sow and Litter: 8 gallons
- Boar: 8 gallons

Again, these are just averages; your animal's water intake may differ depending on the individual situation.

Keeping your pigs properly fed really isn't as difficult as it sounds. And, whenever you're not sure about something, feel free to ask your vet or feed expert. Above all, watch your pigs: they will likely let you know when problems may be occurring. Adjust food intake as necessary, and give them unlimited access to water, and you're well on your way to healthy, happy, and well-fed porkers!

CHAPTER 5

DISEASES AND AILMENTS

N o matter how well your animals are kept, sooner or later you will run into the occasional problem with your pigs. The truth is that there are a number of different health problems that your pigs can develop, so it's important to be informed. This chapter covers just a few of the more common diseases and ailments that can affect your animals. Note that these are only basic descriptions, and should not be used for diagnosis. If you see some of the symptoms listed in this chapter in any of your pigs, make sure to call a qualified farm doctor or veterinarian to come take a look at your animals.

Clostridial Disease

This bacterial infection, which is fatal in most cases, is caused by a bacteria belonging to the class **Clostridia**. These bacteria can enter the body through damage to the skin, muscles and/or underlying tissue. The spores can also lie dormant in the liver for long periods at a time. The course of the disease is usually short, with no outward signs except for a dead animal. This is due to rapid rate of multiplication seen in the toxins the bacteria create, which rapidly kill the host.

There are two known strains of clostridial. The *C. novyi* species can be a problem in outdoor breeds, and is most commonly found in sows. The second is *C. perfringens* type A, which may be milder and more prolonged. It can cause severe diarrhea in piglets, resulting in high mortality. This strain is seen when the bacteria enters the small intestine and establishes itself before colostrum is consumed by the piglet. Animals with this strain are usually infected within the first 24–72 hours of life.

In piglets, early symptoms may include watery diarrhea (**scour**) which has a rotten smell and/or blood color mucus. This diarrhea will usually be seen at either the 0–5 day or 6–21 day mark after birth. There could also be necrosis, where the small intestine sloughs off, and is sometimes seen in the scour.

In weaners and growers both, clostridial disease can cause sudden death, feces hemorrhage, diarrhea, and/or swelling over the muscle masses, which can be both discolored and painful. Gangrene may also result.

Causes include an environment high in bacteria, muscle trauma, fungal poisoning, or the bacteria entering through skin damage.

Salt Poisoning

Usually caused by water deprivation, **salt poisoning** can affect all ages of pigs, and occurs when the normal level of salt becomes toxic due to lack of water. Symptoms include dehydration, lack of appetite, nervous or restless behavior, followed by fits in which the animals will wander as if blind, pressing their heads against the wall. Their snout will usually twitch before convulsions.

There is a high mortality rate with salt poisoning. As a result, make sure that your pigs *always* have access to sufficient water.

Heat Stroke

Symptoms of **heat stroke** include distress, vomiting, red skin, and trembling muscles. Heat stroke is caused by high temperatures, which can be further exacerbated by poor ventilation and high humidity in the pig's shed.

As with any other animal (including humans), pigs affected with heat stroke must be treated immediately. If possible, place the animal in cold water. If the pig affected is too large for your tub or container, spray it with cold water.

To treat heat stroke, place the pig in cold water or spray it with cold water.

Internal Parasites

White ascarids, red stomach worms, and **whip worms** are just a few of the internal parasites found to affect pigs. The eggs of these parasites are often found in the pigs' feces, which leads to an adult infestation.

The sow in particular must be kept safe from any internal parasites, as she will serve as the source of any potential piglet infections.

Symptoms of infection by internal parasites include:
- Loss of condition in sow
- Coccidiosis (in weaners/growers)
- Coughing (in piglets, commonly caused by threadworm)
- Vomiting/blood in feces
- Sloppy or bloody diarrhea
- Paleness
- Visible signs of pain and suffering
- Mortality

Causes of parasite infestation include poor management of living areas, feces accumulation, access to moist/wet areas, wet and dirty floors, and transfer from infected pigs.

Pseudorabies (PR)

Caused by the herpes virus, **pseudorabies** can be found hidden in the nerves of the carrier pig for quite some time before it reactivates. Once in the herd, pseudorabies remains; it's most notable effect is disrupting reproduction. The virus can also survive for up to three weeks outside of the pig. Outbreaks usually happen when strains of the virus infects an unvaccinated herd, and may also infect cattle and dogs.

Symptoms in sows include coughing, fever, abortions/reproduction failure, still births, mummified litters, and weak litters at birth.

Symptoms in piglets include nervousness, sneezing, and coughing; in weaners and growers, symptoms include fever, sneezing, coughing, and pneumonia.

Foot Rot/Bush Foot

An infection of the claw, **foot rot** or **bush rot** usually occurs through penetration of the sole through cracks at the sole/hoof meat, or due to a split hoof. Infection is typically in only one foot (usually a hind foot). When the infection starts in the soft tissue between the claws, it is considered foot rot. The infection can include both superficial and deep infections of the soft tissue.

Symptoms of foot rot include lameness, painful, swollen claw, cracks where the sole/hoof meets, or splitting of the hoof. As it progresses, the claw enlarges, and joint infection/inflammation occurs. Swelling becomes visible, usually around the coronary band. Abscess may form and come to the surface.

In the event of hoof rot symptoms, a trained veterinarian should be consulted; however, foot rot is usually treatable using penicillin, tetracycline, or other antibiotics. Anti-inflammatory medication can help with pain management, provided the sow is not gestating. The damaged hoof wall material may also be removed, preventing the spread of further infection.

Abortion and Embryo/Fetal Loss

In pigs, **fetal losses** usually affect less than 2 percent of sows. Fetal loss can happen anytime from 14–110 days after mating. Symptoms include delivery of a premature litter (with or without mummified piglets), mucus, and blood or pus discharge from the vulva.

Causes include common infections like parasites or the influenza virus. Noninfectious causes include seasonal infertility, poor nutrition, poor hygiene, or a reaction to a vaccination. Moldy feed, contaminated water, and stress can also be a factor.

Piglet Anemia

Piglet anemia is caused by a lack of hemoglobin resulting from a shortage of iron. Piglets are pale, with fluid build-up around the throat as well as internally. There may by some scour (watery diarrhea). Anemia opens the piglet up to a host of other problems, as well. Untreated, anemia is responsible for 10 percent of pre-weaning deaths among piglets.

Piglets can get the iron they need through injections or oral doses of iron, preferably within 18–24 hours of birth. After weaning, piglets will get the iron they need from their feed.

E. Coli (Scours)

Scours refers to what is basically diarrhea in pigs, and is most common in very young piglets. It can cause high mortality/morbidity, and can occur at any age during suckling time. Peak times for scours are before 5 days of age and between 7–14 days of age.

Acute symptoms include piglets huddling together and shivering, or else living in a corner. The area around the tail and rectum may be wet, and there may be vomit. Scours may have a distinctive smell, and be watery to creamy in consistency. As the diarrhea progresses, dehydration, sunken eyes, and leathery skin appear. Before death, piglets may be on their side, frothing at the mouth and paddling.

Early symptoms include similar signs as with acute cases, only less dramatic. There is a lower mortality rate, but symptoms are prolonged, usually occurring at 7–14 days. The diarrhea may be watery to creamy, and its color will be white to yellow.

Symptoms (in weaners) include loss of condition and dehydration, watery diarrhea, sunken eyes, and rapid loss of weight due to dehydration. They may also be found dead with sunken eyes and slightly blue extremities.

However, there may be no symptoms at all; you may just find one of your piglets dead, suddenly.

Causes of scours include:

- Piglets: poor floors, bad drainage, poor hygiene, contamination, drafts, routine use of milk replacer, not enough colostrum, and problems accessing teat
- Weaners: stress, high/low temperatures, poor pen/house hygiene, not enough water, and poor nutrition
- Post-weaning: chills, temperature, poor flooring, or other illnesses
- Growers: diarrhea, swine dysentery, parasites, colitis, salmonellosis, and porcine epidemic diarrhea

As scours can be very serious, contact a veterinarian at the first sign of symptoms for immediate treatment.

In the end, even the best kept animals may end up injured or ill. However, if you pay close attention to your pig's health, their environment, the cleanliness of their shed and pen, their temperature tolerance, their nutrition and the actions of the pig itself, the health risks to your animals (and resultant losses) should be minimal.

Thermo-Neutral Zone

Along with aliments, a brief discussion of the thermo-neutral zone will be valuable in keeping your pigs happy and healthy. The **thermo-neutral zone** refers to the temperature range required for the best productivity of your pigs, in addition to those temperatures which can be considered critical (these vary according to the weight and body condition of your individual animals).

The "limits" or the thermo-neutral zone are defined by the highest and lowest temperatures your pigs can endure and still function properly, without difficulty. These limits are referred to as the **evaporative critical temperature** (ECT) or upper limit; the lower limit is the **lower critical temperature** (LCT), and the **upper critical temperature** (UCT) is the point at which your pigs begin to experience a loss in condition or function.

The UCT is the highest tolerable temperature a pig can experience before serious problems begin. If the temperature of their immediate surroundings rises above the UTC, the pig will become distressed. The UCT also decreases as the pig ages. Temperatures over 80°F (27°C) are undesirable for growers, finishers and breeders. Heat stress (a common problem in hot, dry climates) may be reduced through spray cooling, or providing the pigs with wallowing areas. Evaporation of water from the skin can help reduce excessive body heat. However, as the humidity in the shed rises (if the pig is indoors), this becomes less effective.

Likewise, if the temperature falls below the LTC, a pig will use his/her energy in order to keep warm. Younger pigs/piglets will suffer from the cold, but older pigs will tolerate a lower LTC. Favorable temperatures for newborns are between 80–95°F (27–35°C). If the temperature remains below 60°F (16°C), loss of function can happen quickly. Below 35°F (2°C), fatal chilling occurs within minutes unless the animals are warmed immediately.

CHAPTER 6

BREEDING AND PIGLETS

O n the small homestead or backyard farm, most will choose not to **breed** their own pigs, and will instead purchase a couple of pigs at an auction (remember that pigs are social animals, so you *will* need at least two). Or, better yet, some homesteaders will purchase their new animals from an area farmer who may be selling piglets. These farmers will raise the pigs to slaughter weight, send them off to be slaughtered, and then begin the process again the next year. However, as a few may decide to breed or may be interested in the breeding process, this chapter will briefly discuss what goes into breeding your own pigs.

Is Breeding Right for Me?

As a note, some will discourage backyard farmers from breeding at all, saying that it is too difficult. However, from my own experiences as a pig breeder, I have to differ in that opinion. I don't know if I was simply fortunate in my choice of animals, or if it was because I actually allowed my pigs to be pigs—choosing to let my instincts, the pig's natural instincts, and nature guide me—but I enjoyed the time I spent breeding my own pigs.

However, it is certainly true that breeding isn't for everyone. Remember that at the end of the breeding process, you will (hopefully) have a litter of piglets that you will then need to sell off. You'll also need to decide *when* you want to sell your animals. I chose to sell my piglets at 8 weeks old, to those who were either looking for pigs to raise for food or else looking for new or replacement boars (I had gained a reputation for gentle animals, boars included, which I will go into in more detail later on). I was also fortunate in that I always had more buyers than piglets available, which was good in case someone backed out (which, I have to say, rarely happened). I usually had a waiting list as far as a year in advance for my piglets—but not everyone is so fortunate. If the market for new piglets isn't large enough to support your business, you could find yourself with a quickly-growing litter of new pigs to feed and care for. So, should you decide to choose this route, make sure to check your market first.

Now, you may ask why I decided to sell so early, at the 8-week mark. First of all, it made them more attractive to potential buyers: by then, all of the piglets were weaned and eating on their own. I actually ended up with a better return for myself in the long run, as well. After all, I was getting top dollar (at the time) for an animal that I had put 8–10 weeks of work into, along with maybe 4–7 weeks of food, after which they would leave to go to other farms, at which point I would be back to my two breeders. I decided not to raise the piglets to slaughter weight, as it would have meant more money going into them, more time required, and more buildings needed. In addition, slaughter weight pigs were not always as easy to sell, as you needed to sell to someone looking for meat or breeding. My costs would also need to include transport to the slaughterhouse, which would require the purchase or rent of a trailer. I would need to have the buyer coordinate with the slaughterhouse on what they wanted. To top it off, if the buyer only wanted half a pig, say, I would need to find someone willing to purchase the other half before the animal went in. In the end,

I found that the few extra dollars I might get selling on the hoof just wasn't worth it when I had a lucrative piglet market, with animals in and out within 8–10 weeks.

The Breeding Process

Now, let's talk a bit about the **breeding process**. This is not a comprehensive description; it is just meant to give a very brief idea about the process. If you are seriously considering getting into pig breeding, there are many comprehensive books, papers and websites on the subject. Your local extension office should also have this information available. Your best learning opportunity, however, would be to learn directly from other breeders. If you're fortunate enough to have one in your general area, ask if they will mentor you and let you see some actual breeding sessions, births (if possible; pigs usually need no assistance during birthing, so the actual birth may be hard to catch), and the piglet phase. The breeder may even sell you your first pair, or a bred gilt or sow, if you still feel you want to breed. In the absence of an actual farm, your next best bet will be a few good books and getting on some of the homesteading and pig forums on the Internet.

Should you decide to breed your own pigs, you have a few choices on *how* to breed. Most of these different options stem from where you'll be getting the material needed to inseminate your sow. Some will keep their own boar, others will borrow or rent a boar (which can be a difficult task, especially if there are no other pig farmers around), while others will turn to artificial insemination (AI). Finally, you can also purchase an already-bred sow.

There are a few terms that you need to know when it comes to breeding pigs:

- **Bred**: A pregnant female pig. Depending on the stage of life, the animal may be referred to as a bred gilt or a bred sow.
- **Open**: A female that is not bred (at the time).
- **Farrow**: This term may be used in one of two ways. As a verb, farrow or farrowing means to give birth, or giving birth. As a noun, it refers to a litter of newborns.

When Should I Breed?

The **breeding ages** for boars and gilts depend on the individual animal and breed, but the average breeding age for boars to begin is 28–30 weeks, while gilts will mature at 18–24 week. However, many small keepers may prefer to have the first breeding at 8–12 months (this was the age range of my gilt for her first breeding).

Gilts are usually mated on the first day of **heat**, while sows are mated on the second. How can you tell when your female is in heat? Once you have a few cycles under your belt, it gets easier to tell. But if you are new at breeding, signs of your female being in heat include a swollen vulva, frequent urination, restlessness, mucus or bloody discharge, and a twitching tail. Although your sow will probably not have every single one of these signs, she will usually have at least one. Note that if you *do* decide to keep a boar, you may need to reinforce their pen area with electric fencing; if you have a sow come into heat, the boar may cause damage to his pen (and hers) trying to get to her.

Once your female is in heat, she will be in a **standing heat** for 36–60 hours (this is the period during which she is willing to breed). At this point, you will either need to take her to the boar—it is best to take the sow to the boar, rather than the boar to the sow, as this will lessen the potential of having the boar become excited from being moved, which in turn, could make him more difficult to handle—or else have the sow artificially inseminated. Gestation for pigs is 113 days (or 3 months, 3 weeks, and 3 days) with the countdown starting from the time the boar and sow are together (the day of insemination). The estrous cycle is 21 days, so if your female does not take the first time, you can re-breed every 3 weeks until she does (if the female did not take, she will have another heat cycle).

Even if your animal did take, know that the female will not show signs of pregnancy immediately. It will take 15–20 days for the fertilized eggs to implant into the uterus. In pigs, the embryos will only remain viable if **more than four embryos successfully implant,** in which case, the sow or gilt's body will release enough estrogen to maintain the pregnancy. Otherwise, the female's body will reabsorb the embryos and her normal estrous cycles will begin once again.

For the first 30 days, the fetal skeletons are almost like body tissue; if there is a problem, the sow's body can reabsorb the fetuses. After 30 days, the bones in the fetal skeleton will begin to calcify and harden. Should any problems arise from 35 days and forward, it could lead to delivery of mummified babies. At 60 days, the fetuses have developed most of their **immunocompetence** (ability to develop an immune response) using the mother's antibodies. From this point on, the fetuses will grow and finalize the development of their systems, brains and other organs.

Birth and Piglets

Just like humans, counting the days does not guarantee the delivery date, but it does give you a good idea. The female will also usually give some signs that delivery is imminent:

About four days before delivery, the female may have a swollen or red vulva. Her teats may also be leaking; if so, the first birth may be up to 48 hours away. The whiter the clear fluid turns, the closer she is to giving birth. Pigs will also build nests in anticipation of birthing; when you see them starting to construct a nest, delivery is usually about 24 hours away. I would also sometimes notice a vaginal discharge 24–48 hours before my sow gave birth.

The female will give birth in a lying position, not standing. The piglets are usually born head/snout first. Importantly, within 24 hours of birth piglets are usually given **iron**, to prevent anemia. This can be through injection or oral drops (I've always used the oral drops, as it is much easier to do and causes little stress to the animals). It is also worth mentioning that the sow and her piglets can recognize each other through olfactory and vocal stimulation.

Nursing

The piglets are born with teeth—*sharp* teeth. Piglets may begin to suck right after birth or very soon after. You'll soon discover that the behavior of the piglets and their mother during nursing is somewhat complex, with the piglets jockeying for position at the teats and the mother responding to her offspring in a series of specific grunts. When the piglet finds his or her spot, it will massage the teat with the snout. The sow will, in turn, grunt at slow, regular intervals. These series of grunts will vary in magnitude, frequency, and tone. These variations will signal the **stages of nursing** to the  piglets. Even at this early stage, you'll begin to see hierarchies forming: the piglets will develop a teat order, which will usually remain for as long as they continue to nurse.

After the **first stage** of suckling, which comprises of the piglets finding their place and massaging their chosen teat, milk should begin to flow—this is **stage two. Stage three consists** of the piglet sucking, with mouth movements at a rate of about one suckle per second. At this time, the grunting of the

sow increases for less than half of a minute. At this phase, the peak of the mother's grunting will coincide with the release of **oxytocin** (a nonapeptide hormone) from the pituitary into the blood stream, triggering the release of milk.

Stage four is a 10–20 second period of main milk flow. The piglets will withdraw a bit from the teat before beginning a rapid sucking. The sow will grunt rapidly in a low tone, sometimes in a quick successions of three or four grunts.

Finally, the milk flow stops, as do the grunts. The piglets may dash around the teats and resume a slow suck or nose the udder, and will massage and suckle the teats after the flow finishes. This will help the sow regulate the flow from each teat in the future. The piglets will nurse approximately once every 60 minutes. Each time, the sow will require the piglets' stimulation behavior before she can release their milk.

Weaning

Weaning usually begins between 4–8 weeks of age. This is usually a stressful time for the piglets, as many times it is a **forced wean,** where the piglets are forcibly separated from their mother. This is also stressful for the mother as well; she will most likely look for her litter and be a bit uncomfortable until she dries up.

Some have tried **self-weaning**, meaning letting the sow and her piglets wean naturally; however, most report little to no success with this natural method. I was fortunate enough to have natural weaning work for me, and I do swear by the method. My piglets were weaned completely by 6–7 weeks of age using this method, without fail. It was also less stressful on the piglets, the sow—and me! However, I again remind you that many who *have* tried this method have not had luck with it, so should you decide to give natural weaning a try, be prepared for the possibility that you will still need to remove the piglets from the mother's pen to complete the weaning process.

Some farmers wean their piglets at 3–4 weeks of age. In my personal opinion, unless the piglets decide to begin weaning themselves at this age (which I did have a few do), this is a bit too young to force wean. Reason being, it is stressful for very young piglets to go from a high lactose diet to a diet of complex starch (unless you plan to develop your own balanced diet for them, although it will still be stressful, whatever you do), and **creep feeding** (offering regular food to still nursing animals) is usually not even touched by the piglets until 3 weeks of age. I have occasionally seen piglets attempt to nibble on mom's food before 3 weeks of age, but it was never forced, and seemed more like curiosity than hunger. The piglet's digestive tract is still under-developed at this point; the piglet needs to adjust to any new diet, and they will have a limited ability to produce antibodies at this time (as they were previously coming from their mother's milk). There is also the social adjustment to consider when weaning—going from the sow to an entirely new environment, which includes new pen mates.

The decision of when and how to wean your piglets is yours, and yours alone, but for my two cents, forced weaning much earlier than four weeks is still a bit much for a piglet at this very young age, unless it is absolutely necessary (to address the animal's health needs, for example, or with an aggressive parent).

Orphans

It is an unfortunate fact that sometimes you may end up with **orphan pigs**, pigs that either aren't or can't get milk from their mother. This is usually due to a number of factors: there are too many piglets for the sow to handle; an injured or ill piglet; the sow has rejected piglets; or else the sow has gone into lactation failure (Mastitis-Metritis-Agalactia). Signs of **lactation failure** include: hungry piglets; a depressed and/or feverish sow; the sow doesn't want to eat or get up; and/or constipation. When this happens, call your vet immediately, as there are specific drugs (including penicillin along with electrolytes) which will get her going again (usually within four days), at which time her milk should begin flowing once again.

Should you find yourself with an orphaned piglet, either temporarily or permanently, you can feed the orphan from a pan or a bottle (a baby bottle for humans works well, although you may need to enlarge the hole in the nipple a bit). A little one may warrant **bottle feeding** for a while, while a piglet that is a few weeks old may be able to use a bowl. If your orphaned piglet has to be taken at birth, make sure that it gets colostrum within the first 24 hours. **Colostrum** is basically the "first milk" that the piglet receives from the teat, which contains antibodies that will help protect the newborn from disease. It is critical that the newborn piglets receive colostrum within this 24-hour time period. You may purchase colostrum replacement at feed stores; or, if you know someone who raises goats or dairy cattle and freezes colostrum for emergencies, perhaps you can obtain some from them.

You will also need a **milk replacement** for the orphaned piglet. Although goat's milk is preferable, you can also use cow's milk. A good milk replacer may also be purchased at most feed stores; talk with your feed person if you're not sure. You will need 24–28 percent protein and 8–10 percent fat. There are also a number of recipes available for homemade milk replacement. Two recipes are as follows:

1 quart milk
1 raw egg
1 pint half-and-half
4 cc neomycin

Mix ingredients together and heat until warm. You can use any type of container or heat source, provided the milk replacement doesn't get too hot.

2 pints milk
½ cup half-and-half
¼ cup dark corn syrup
1 egg

Mix ingredients together and heat until warm. You can use any type of container or heat source, provided the milk replacement doesn't get too hot.

For extra information, there are also a number of feed, growth and weight charts available online and in books to help guide you through the various stages and needs of your animals.

Castrating
Of course, your litter will contain both boars and sows. Most will choose to **castrate** their young boars, unless they plan on selling the boar for breeding purposes. There are many who choose not to castrate, as well; they claim they do not see any difference in the taste of the pork between castrated and intact boars. I myself

do not castrate, but this was primarily because people bought my animals not only to raise for food, but as breeding stock as well.

If you *are* castrating, you can do it right at home. However, this is not for the squeamish, and you will need help. The first few times you do it, you should have someone who knows what they are doing there to guide you. To give you an idea of the process, I have included the basic steps. Again, do not try this on your own unless you have an experienced person with you.

1. Restrain the piglet. It's best if you have someone else to hold the pig.
2. Using a *clean* scalpel, cut the skin on the scrotum.
3. Pull the testes away from the body.
4. Cut or sever the spermatic cord with the scalpel.
5. Spray the open wound with disinfectant (purchased from a vet or feed store).

The piglet will experience pain and distress (and you probably will too, on your first few), but the pain will usually subside within hours of the procedure. The piglet will be screaming and obviously trying to fight, so your helper must be able to have a firm grip and endure the loud screams that will likely be ringing in their ears for the rest of the day!

As a final note, average losses at pre-weaning age are 11 percent, with at least half of those losses occurring within the first three days after birth. The cause can be anything from the piglet being too small to an accidental crushing by the sow. No matter how vigilant you are with your litters, at some point there will be some losses. The best thing is to understand that it *does* happen, see if there is anything that you can or need to correct, chalk it up to experience, and move forward with the remaining animals.

CHAPTER 7

CULLING AND LIVESTOCK SLAUGHTER

· ·

Culling

It's a common misconception that culling and slaughtering an animal means the same thing. And this is understandable: at times, the process of **culling** (which means removing an animal from the herd) may entail slaughtering the animal. This is often the case when the animals being culled are carrying some kind of disease or ailment, in which case they are slaughtered to protect the rest of the herd. Frequently, however, culling does not require that the animal be slaughtered—it need only be removed from the herd.

Reasons why you may need to cull an animal include:

- the animal does not conform to **breed standards** (this is especially important when breeding purebred animals for registration)
- too many animals (the herd is too large)
- old age

- sickness that is incurable, or will otherwise render the animal useless for your needs
- the animal is not performing as needed

When selecting animals to be culled, you must be certain of your decision, especially if you are selling or slaughtering the animal. Removal will be permanent, so you need to make sure that you are making the correct choices. If you are uncertain when it comes time for your first culling, try to find an experienced person or even a farm/large animal veterinarian with experience in pigs to help guide you in what to look for when it comes to deciding to cull or to keep. Once you have decided which animals to cull, you will next have to decide what you will be doing with the animals. If you don't want to slaughter them, and you don't have ready buyers for your culls, you can always take them to a livestock auction.

Slaughtering

Although far from the most pleasant job on the farm, **slaughtering** is necessary if you want to turn your pigs into pork. Thankfully, you have a few options when it comes to slaughtering 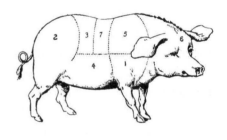 livestock. If you have a **slaughterhouse** in the area, you may send your animal there. If you are concerned as to how your animals will be treated, don't be afraid to see if you can arrange a visit to the slaughterhouse facilities. Talk with the people who work there and see how they handle and treat the animals. If it is up to your standards, go ahead and make an appointment for your animals to come in. If you really don't like what you see, thank your host and look at other options.

If you want to try **home slaughtering** (which is much less stressful for the animal) but do not want to do it yourself, look for an **experienced person** who will come to your farm and do it for you. Of course, there will most likely be a charge for this service; however, if you are set on a home slaughter, but the thought of doing it yourself doesn't set well, this option may be best for you.

The third option is a home slaughter where you are doing the work yourself. If you have never done a home slaughter yourself before, you will definitely want an experienced person there with you to teach you and guide you through the process.

Using Every Part of the Pig

The expression, "Everything on a pig can be used except the squeak," is not just a wives tale; it is a fact. Every part of the pig can be used, from the head (for headcheese, as well as other uses) to the jowls (for smoking or curing—delicious in collard greens) to the intestines (for natural sausage casings) and the fat (rendered for lard or soap). If you look hard enough, you'll find that every part of the pig has its uses.

Prepping for Slaughter

The following are the basic steps for home slaughter, to help give you an idea of what is involved. (This should *not* be considered a complete guide for an actual slaughtering. As this is most likely your first slaughter, make sure to have someone with you who knows exactly what they are doing and can teach you properly.)

For home slaughter, the best time of year is late fall to early winter. This allows the carcass to hang and cool safely for 24 hours or more before sectioning, freezing, and/or curing. If it is cold enough, the carcass can hang from (strong) rafters in an unheated shed or barn until the flesh cools to between 33–35°F. However, if the slaughtering is to be done at home in a warmer time of year (or if you have a unseasonably warm late fall or early winter), you will either need to cool the carcass down with ice or a cold brine, or else rent a cold storage locker large enough to hold the carcass.

You should have at least one person assisting you in a home slaughter—two, if possible, including the experienced person there to teach you the correct and humane way to carry out the slaughter (as well as the correct handling of the carcass). No later than 24 hours before the slaughtering day, you should remove the hog to be slaughtered to its own pen. This will give it time to calm down, and prevent any bumping and excitement from the other pigs that could affect the animal. You should also withhold food from the pig for at least 12 hours prior to slaughter, 24 hours if possible. The pig *should*, however, have full access to all the water it wants. This will help to purge or clean out the digestive system. (Even if you only have two animals and are planning on slaughtering both of them, it will still be a good idea to separate them.) If they will tolerate it, you can wash the animal(s) if you so choose. If the animal does not want to be washed or hosed down, just leave it alone and let them calm down.

It is not a good idea to slaughter a gilt or sow that is in heat, as it can affect the flavor of the meat. Instead, wait for their period of heat to be over. It is also said that intact boars should not be done for the same reason; however, as we have discussed in previous chapters, people are experimenting and have found that this is not necessarily true, especially in young boars. In the end, the choice is up to you.

Once you have your pigs taken care of, it is time to get your equipment together. A basic list of slaughter equipment includes:

- A rifle, at least .22 caliber
- Two good, sharp knives: one at least 6 inches for sticking (bleeding) and a smaller knife (2–3 inches) for gutting
- Sharpening stone (to keep knives razor sharp)
- Hog or bell scraper (for removing hair and **scurf**—the thin membrane on the hide)
- Hot water thermometer
- Large vat, for scalding (large enough to dunk the entire carcass, if possible)
- Food-grade bucket, to catch blood (if you are keeping the blood for culinary use)
- Some type of table, large and sturdy enough to hold the carcass for scraping
- Tub, to catch the viscera or innards when gutting (these should be caught in a large, *clean* plastic bag or food grade vat if keeping the organs for culinary use)
- Come-along or some type of pulley system to hoist the carcass up (this will help immensely when doing the scalding and gutting)

A few days before slaughtering, you should have your equipment organized and begin setting up your workstations. For some items, you may want to set up a day or so before; other things can be set up early that morning. Some recommend that the workstations be set up close to where the pig is penned to make it easier on both you and the animal. By keeping the station close by, you avoid overstressing the animal and you avoid having to move the animal very far. Your stations should include the slaughter/gutting area, dipping/scraping area, and your processing area. When setting up your work areas, make sure that you have access to plenty of water.

Slaughtering Your Pigs

On the day of the slaughtering, you should have everything set up and ready to go *before* you get the hog. Where the deed is to be done is up to you; whether in the pen, or in the work area by the come-along. Again, do not agitate the pig in any way; not only is it cruel to the animal, it can affect the quality of the meat.

The best and most humane way to initially put the pig down is with a .22 gunshot, point blank to the head. The general consensus is to draw an imaginary X between the animal's ears and eyes, with the center in the approximate area of the brain. Do *not* put the muzzle of the gun against the animal's forehead. Not only will this scare the pig (again, you're aiming to cause no stress to the animal), the gun itself may malfunction. When you go to shoot the hog, if you are afraid it will get skittish or you just don't want them to see it coming, put down a small bowl of feed for the animal and wait until they begin to eat. If you are not sure as to how to make the shot, let an experienced person do this part; the pig should be brought down with a single shot whenever possible.

Once the pig is down, either dead (hopefully) or else rendered totally unconscious, turn the animal on its back (if you want to bleed it on the ground) or hoist it into the air with the come-along. Cut and stick in the throat area, and allow the animal to **bleed** out (note that hanging is much easier, and is the better method for this process). If you are planning to save the blood, your food-safe bucket should be placed accordingly in order to catch it. The bleeding phase is very important; if the tissues retain too much blood, the meat will be tainted.

The bleeding process can take

Photo by Pietro Pensa under the Creative Commons Attribution License 2.0.

several minutes during which time, even if the animal *is* dead, you will witness some twitching and motion. A sufficient bleed out will render at least 4 quarts of blood. If you plan to keep the blood for culinary use, once the process is complete you will need to **stabilize** the blood so it doesn't clot and turn to gel. To do so, add ½ cup sea salt to the bucket of blood and stir for approximately 30 seconds, and then scoop/strain off the white foam. The blood is now stable to store in the refrigerator for weeks without turning.

After the bleeding comes the **scalding**. The carcass is lowered in head first, rocked and soaked for 3–6 minutes, and then hoisted onto the table to begin removing the hair and scurf with the hog scraper. Scalding allows the hair and scurf to be scraped off without losing any fat or skin. If all the hair does not remove the first time, the pig may be dipped a second time for another 3–6 minutes.

The **scalding vat** will need to be heated at least an hour before slaughter, as the water temperature needs to reach 145°F, but not *exceed* 160°F, as this risks cooking the pig. To prepare your scalding vat, fill the vat two-thirds of the way with water and add some lime. The easiest way to heat the water is using a wood fire under the vat (or cauldron); however, some have rigged up a propane-based source for the fire and noted good results.

After the scalding and scraping have been completed, the hog should be **gutted** (at this point, the hog should be hanging if it wasn't already). There should be a tub underneath to catch the innards, especially if you plan to use them for culinary purposes. The process of removing the organs is known as **evisceration**. When gutting the hog, it is best to use a 2–3 inch blade so as not to spill rumen into the abdominal cavity, which can taint and spoil the meat.

After the animal has been gutted, the next step is **cooling** the carcass, which we discussed earlier in this chapter. Once chilled to the proper temperature, which can take a few days, the carcass is ready to process for freezing and/or curing.

What Age Should I Slaughter?

If you are wondering what the correct slaughtering age is, there really isn't any. Although the prime slaughter *size* is 240–260 pounds, there is also what's known as a **suckling pig**, which is a piglet fed on "mother's milk" and usually slaughtered at 2–6 weeks (although some consider piglets to be suckling  pigs up to 3 months of age). Usually roasted whole and very tender, the suckling pig is normally used for special occasions and special gatherings, while **fattening pigs** are usually done at 4 months to a year in age, and are usually primed for pork and bacon.

No matter what method of slaughtering you choose, remember that the main objective is a humane slaughtering, with the pig kept calm and basically stress-free. If you plan on home slaughtering, remember that either the temperatures must be right or you must have an appropriate cooling spot and at least the minimal amount of proper equipment is essential, including razor sharp knives and extra help to make the task as easy as possible. And, if you are new to home slaughtering, I can't stress enough how important it is to get yourself a good teacher who will guide you through the entire process.

CHAPTER 8

PIGS AS PETS AND OTHER USES

Pigs as Pets

Though you might not think it to look at them, pigs make wonderful **pets**. And, while you can certainly purchase potbellied pigs, teacup pigs, and other mini pigs to serve as pets, a regular old farm pig also makes for a loveable companion as well. Although pet farm pigs usually get relegated to an outdoor pen due to their size, they are still just as loveable, just as fun and just as responsive to their owners as their smaller counterparts.

When raised as pets, farm pigs can be just as gentle as their miniature counterparts. In fact, I found my farm pigs to have a better temperament than my potbellied pig (who was litterbox trained *and* lived in the house!) The key to raising a farm pig as a pet is to **get them while they're still young and work with them on a daily basis**, just as you would with any other pet. Get them used to being touched, walking on a lead, and being around people. My two breeders were raised as pets, as I knew I was going to have to deal with them on my own. It was certainly worth the extra time; by the time I was done, I could walk my 1000-pound breeding boar on a lead!

So, if you find that you have become hopelessly attached to one of the pigs that was intended for your freezer, don't be afraid to make them into a pet. As long as you have the time to put into the animal and keep in mind how big your pet may get, go ahead and take the plunge. Not only will it be the most fun pet you will ever have, it may well be the most intelligent.

Other Uses for Pigs

Besides the uses we've discussed for pigs (for breeding, for meat, as pets, etc.), there are a number of other potential uses for your talented swine.

Due to their keen sense of smell, some pigs (mostly smaller breeds, such as potbellied pigs) have been trained for **drug and explosives sniffing**. Although there's little chance that pigs will be putting drug-sniffing dogs out of business any time soon, the pig *has* been making its mark.

Another job the pig has been given, again due to its sense of smell, is that of **truffle hunter** or **truffle hog**. Truffle hogs can locate truffles as deep as 3 feet underground. They are trained to hunt on a leash, walking with their owner. It is said that the use of pigs to hunt truffles dates back to the Roman Empire;

A truffle hog is trained to locate truffles underground. Photo by Robert Vayssie under the Creative Commons Attribution License 2.0.

however, the best evidence indicates their use began in the 15th century.

It is believed that the pigs can sniff out these truffles, which are found hidden underground, due to the fungi having a smell similar to a boar's sex hormones; as a result, the pigs used for this job are primarily female,

Although truffle hogs are still in use (and a good truffle hog is still a very valuable animal), some truffle hunters are switching over to the use of dogs. Their reasoning is sound: the truffle pig's taste for the expensive truffle ends up with the animal eating quite a bit of what she finds. On the other hand, traditionalists insist that the pig has a much more sensitive nose than a dog, and that her taste for truffles makes her a better, more loyal hunter.

It should be noted, however, that Italy banned the use of pigs in truffle hunting in 1985, stating that the animal damages the truffle's mycelia (vegetative part of the fungus) as they dig, which had ended up in a reduced production rate for quite some time.

Perhaps the most unusual use of the pig can be found in the making of a rare Tuscan cheese called **Porcorino** (Porcherino), which is created from pigs' milk. Made in a southern town in Tuscany and produced by a single family, the cheese-making process is a closely guarded secret. Created in small batches intended for local use, the cheese is formed into small, firm rounds, 1–2 inches in diameter. It has been produced for hundreds, perhaps even thousands of years.

Although pigs' milk is very high in butterfat (at 8½ percent), pigs can only produce an average of 13 pounds of milk per day—much less than a cow. And it goes without saying that milking a pig is tough; between the number of teats, how quickly each needs to be done and their relative size, it's a tricky process which requires a significant amount of trust between animal and human. However, those who have been fortunate enough to have the cheese say that it is a delicious experience.

As you can see, the pig has far more uses than just comprising your main meal! My hope is that somewhere along the way in your porcine adventures, you will be fortunate enough to experience everything the pig has to offer!

FINAL WORDS

O f all the animals I've raised, I think I found the most enjoyment raising pigs. I've had farm pigs as well as a potbellied pig who was litterbox trained, leash trained, *and* liked to go camping with me and my Australian Shepherd, Cheyenne. And my farm pigs were no less endearing!

Originally, I started with a breeding pair of Yorkshire mixes that I got when they were about 4 months of age. Because I knew that I was going to have to handle these animals myself, I thought of them—and treated them—as pets. I spent time with them every day, often two or three times a day. I would sit in the straw and play with them; I would work with them so that they were used to being touched on every part of their body (so that, if it was ever necessary, a vet could come in and do what needed to be done with them without having to worry about problems). As an added bonus, by spending so much time with my pigs I got to know their personalities, inside and out. Once I was sure that they'd become accustomed to me, I would also bring friends and relatives back to the pen to socialize with the pair. As a result, they were fine with someone else having to come and care for them when I was away.

I was also able to do something with my pigs that I was told couldn't be done: I kept the boar and the sow together at all times, even when the piglets came. In fact, the boar absolutely loved the piglets; it wasn't unusual to see the big boy boar asleep outdoors with a couple little ones sleeping on top of him. He would even sit with the sow when she gave birth, and as each piglet was born, he would go over, nudge it a little and grunt a little greeting, before returning to the sow. If I happened to be out there, he would also run over to me, grunting away as if he was telling me about each new baby, before going back to her. They raised the piglets together, and because of the trust that I had built up with the pair, I was able to handle the babies as needed with no problem, which made it much easier to give the piglets their squirt of iron.

As I did with the adults, I also spent time each day playing with the piglets, starting 24 hours after their birth. This way, they would be easier for their prospective new owners to handle. When the time came for each piglet to leave, I instructed the buyer to bring an animal carrier, not a box or bag. Because the babies were used to me, I would take the carrier into the shed and sit in the straw and wait for the selected piglet to come over. I would then take a little treat and put it in the carrier when said piglet was in front of it. The piglet would go in and I would close the door. When the piglet turned around, it got another treat. No stress, no screaming, no angry mom and dad, and it normally took no longer than 5–10 minutes. It especially amazed those who had purchased piglets from others in the past. But again, that's the goal: no stress if at all possible.

According to the "experts," there was no way I could ever raise my pigs as a family; supposedly, the boar would kill them. These same "experts" also insisted that I couldn't let my chickens anywhere near the pigs, as they were sure to be killed. I guess someone should have told the chickens and pigs that! At almost every meal, the pigs would have at least a few birds join them at the food bowl, sometimes squatting right underneath the pigs' heads as they all ate. And, in all of the years I raised pigs, I had only a few losses of piglets, and only lost one little one due to crushing by an adult. It took a lot of time in the beginning, and my first littler didn't come until the adults were about a year old, but for me it was worth it.

I was also warned that, by never separating the pair, I risked breeding my sow to death. This could not have been farther from the truth, at least in my experience. Suzie had definite control over that situation. It was almost comical; if Ollie wanted to mount her, but Suzie had no interest, she would back herself up into the corner of the shed or fence and sit there defiantly. If he tried to come near, he would get a warning grunt. If he still didn't listen, she would snap (though she never tried to bite him), still sitting. Eventually, Ollie would give up and walk away. By relying on Suzie's instincts and allowing her to breed when she wanted to, I got one litter a year, like clockwork (which was good enough for my needs), usually around late spring/early summer.

I chose to practice self-weaning or natural weaning, which worked very well for me; my piglets were always weaned no later than 6–7 weeks of age, sometimes earlier. Without fail, my piglets were ready to go at 8 weeks, and without the stress of regular weaning.

I had healthy animals, with piglets always sold out. I also had repeat customers. Someone who had purchased a young boar from me to use for breeding saw me over a year later, and told me that the boar he purchased from me was the first boar he owned that he could actually trust. That was when I really knew that all the time I put into my animals does pay off.

In conclusion, no matter what your reasons are for raising pigs, the better you treat them, the better it will go for both of you. Whether you're looking at breeding or just keeping a couple for meat each year, pigs are like any other animal—you will get out of them exactly what you put into them. You will find that the pig is both the smartest and cleanest animal on the farm.

And, if you're not careful, they're the easiest to fall in love with.

Enjoy!

FACTS AND TRIVIA

- Wall Street (the financial district in New York City) got its name from a wall built by the Dutch in the 17th century. The wall was created not only to protect them from invasion, but to keep their pigs confined to the northern edge of the colony (though this is only a theory!).

- Because pigs were considered to be unclean by the Jewish people (pigs were known to eat waste) pork was banned by the Jewish faith around 1000 bce.

- Have you heard about the Chinese Pig Toilet? Although it may sound a bit disgusting, pig toilets (or pig sty latrines) were once quite common in rural China. The pig toilet was an outhouse that was mounted over a pig sty. A hole or chute of some type connected the two, allowing the

Clay model of a pig toilet. Photo by John Hill under the Creative Commons Attribution License 2.0.

unfortunate hogs to consume the human waste. Although now (thankfully) discouraged by Chinese authorities, pig toilets could still be found in some remote areas of northern China as late as 2005! And China isn't the only place where this was practiced: Goa and Kerala (both in India) still saw a part of their populations using pig toilets as late as 2003. Pig toilets are also known to still be used on the South Korean volcanic island of Jejudo.

- Some who home slaughter still practice slaughtering by a full moon. It is believed that the animal will be easier to bleed, and will result in a more tender meat. This is due to the fact that it is said that the meat will not contract as much when cooked.

- As of 2013, there were 977.3 million pigs worldwide, with China, the United States, and Brazil having the highest population of porkers.

- Pigs possess nicotinic acetylcholine receptor mutations which protect them from snake venom. This is a protective adaptation that pigs share with the mongoose, honey badger, and hedgehog. These pocket mutations prevent the a-neurotoxins, which make snake venom so dangerous, from binding to the cells of the pig's body.

- Although pigs also possess apocrine and eccrine sweat glands (with the latter in the snout and dorsal nasal), they do not use them; unlike most animals, panting usually doesn't help pigs to cool off. This is why pigs wallow in mud (which also helps with sun and parasites) or enjoy a nice water bath.

- Pigs use olfactory stimuli (their sense of smell) to identify other pigs. They also have an alarm stimulus (transmitted through auditory senses and pheromones), and an auditory stimulus for social communication. They have excellent hearing and can localize sound simply by moving their head.

- Adult pigs have 44 teeth. They use the rear ones for crushing, while the canine teeth form tusks (on boars), which continue to grow. Boars keep their tusks sharp by grinding against each other.

- Pigskin is one of the toughest hides that we use; however, the skin on a live pig is actually quite sensitive to injury and temperature.
- Pigs have panoramic 310° vision and binocular vision of 35–50°. While it is uncertain whether or not pigs can see color, the presence of cone cells in their retina suggests they may have at least some color sense.

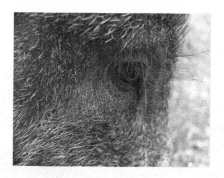

RECIPES

· · · · · · · · · · · · · · · ·

Pork Roast with Cayenne Apple Glaze

2 boneless pork loin roasts
¼ cup Dijon mustard
6 ounces apple cider
2 tablespoons cider vinegar
2 tablespoons brown sugar
2 tablespoons Worcestershire
  sauce

1 tablespoon cayenne pepper
1 tablespoon salt
1 tablespoon black pepper
1 tablespoon smoked paprika

Combine dry ingredients. Set aside.

Spread mustard as needed on one end of each roast. Attach mustard ends of roasts together. Spread remaining mustard all over roasts. Apply dry powder mixture and bake at 375°F until internal temperature reaches 135°F.

Combine remaining ingredients and brush on pork. Bake at 375°F until internal temperature reaches 145°F. Let rest for 10 minutes before serving.

Pork Pot Pie

2 premade pie crusts (may also
use homemade pastry crust)
⅓ cup butter
⅓ cup onion, chopped
⅓ cup all-purpose flour
½ teaspoon salt

¼ teaspoon pepper
1½ cups pork broth
⅔ cup milk
3 cups pork, cooked
1 cup frozen corn, thawed
1 cup frozen peas, thawed

Preheat oven to 425°F. Prepare pie crusts as directed on package for two-crust pie using 9-inch pie pan.

In a medium saucepan, melt butter over medium heat. Add the onion, cooking 2 minutes or until tender. Stir in flour, salt, and pepper until well-blended. Gradually stir in broth and milk; cook, stirring constantly, until bubbly and thickened. Add pork, corn, and peas; remove from heat.

Spoon pork mixture into crust-lined pan. Top with second crust and flute; cut three slits on top. Bake for 30–40 minutes or until crust is golden-brown. Let stand 5 minutes before serving.

New Mexico Pork Tamales

Makes about 50 to 75 tamales

Meat Filling
1 cup cooked pork, shredded
1 cup cooked chicken, shredded
1 (15-ounce) can beanless chili con carne
2 tablespoons cooking oil

1 tablespoon chili powder
1 tablespoon garlic powder
2 teaspoons ground cumin¼ teaspoon of salt
⅛ teaspoon freshly ground black pepper

Wrappers
1 (8-ounce) package of corn husks (contains about 100 husks)

Masa Mixture
2 pounds Masa flour (can be found in most grocery stores)
3 tablespoons paprika
3 tablespoons chili powder
3 tablespoons garlic powder
1 tablespoon ground cumin

2 tablespoons salt
2 cups solid vegetable shortening
8 cups warm chicken stock (or substitute 8 cups water and 8 chicken bouillon cubes)

Meat Filling
Stir ingredients together and refrigerate until needed.

Wrappers
Soak corn husks in warm water for at least two hours. A dinner plate may have to be used to hold the corn husks under water.

Masa Mixture
Mix dry Masa flour with paprika, chili powder, garlic powder, cumin and salt. Mix well for even distribution of spices. Cut

shortening into the Masa mixture with a pastry cutter. Mix until mixture resembles the texture of tiny peas.

Slowly add chicken broth, one cup at a time. Mix well using electric mixer (mixture should resemble consistency of peanut butter). Continue mixing with electric mixer to incorporate air into batter. Mixture is ready to use when 1 teaspoon of batter floats on water.

Filling and Cooking Tamales

Use one corn husk per tamale. Lay flat on table. The corn husk is roughly rectangle shaped. Spread ½ cup of Masa mixture on entire upward facing corn husk, leaving a 1-inch (2.5cm) gap along the long edge and narrow end of the corn husk. Spread 1 tablespoon of meat mixture, in a narrow band in the middle, down the length of the masa spread corn husk.

Roll/fold the husk along the wide edge, touching the edges of the masa together, with the meat mixture ending up in the middle of the masa. The part of the husk with the 1-inch gap with no masa should be rolled around the outside of the husk. Fold about 1 inch of end of rolled husk, upward along the body of the husk (the narrow end with 1-inch masa gap). You should end up with something resembling a cornucopia, with one end folded over and one end open.

Stand all of the rolled tamales on end in a steaming colander with the open end of the tamales facing upward. Steam the tamales over boiling water for 2 hours. Place a lid over the steaming colander. Keep the tamales above the boiling water; don't allow them to stand in it.

At the end of the cooking period, open one tamale to make sure the masa is cooked, and not doughy. Allow tamales to rest 30 minutes before serving (this will help to keep the masa from sticking to the corn husk).

Wiejska

The most popular form of Kielbasa, or Polish sausage, in the United States.

2½ pounds lean boneless pork
1 teaspoon salt
1 teaspoon pepper
1 teaspoon marjoram

2–3 cloves garlic, crushed
1 tablespoon mustard seed¼
cup crushed ice
Sausage casing (natural is best)

Cut the pork into small chunks. Grind with the seasonings and ice (the purpose of the ice is to keep the fat in the pork solid, which is very important in the final texture of the sausage). Mix this well.

Stuff the pork mixture into the casing. Smoke in an outside smoker (following the manufacturer's directions), or else place the sausage in a casserole dish, cover it with water and bake at 350° F until the water is absorbed, about 1½–2 hours. This makes about 2 pounds.

When making sausage by hand, tie a knot about 3 inches (76 mm) from one end of a cleaned sausage casing and fix the open end over the spout of a wide-based funnel, easing most of the casing up onto the spout. Spoon the mixture into the funnel and push it through into the casing with your fingers. Knot the end and roll the sausage gently on a firm surface to distribute the filling evenly.

Warning: This recipe is for fresh sausage. As such, it must be fully cooked *before* eating in order to prevent food poisoning. Alternatively, this sausage can be prepared as a cured meat.

Baked Pork Chops

3 or 4 bone-in pork chops,
  ¼-inch thick
A medium apple, cored and
  sliced thin
A medium onion, sliced thin

1 tablespoon honey
1 tablespoon caraway seed
Garlic salt, black pepper (fresh
  ground if possible) and
  Dijon mustard, to taste

Preheat oven to 350°F.

Layer the sliced onion and apple in a baking dish that can be covered; drizzle honey and half the caraway over the apple and onion slices. Season the pork with garlic salt and pepper; spread top of meat with Dijon. Sprinkle remainder of caraway. Cover and bake for one hour.

Mofongo: Mashed Plantains and Pork Rinds

A popular Caribbean dish originating in Puerto Rico and later adapted to Dominican cuisine. It is made from fried green plantains, seasoned with garlic, olive oil and pork cracklings, then mashed, and usually served with a fried meat and a chicken broth soup.

3 plantains, very green
½ pound pork rinds (also known as chicharrones or pork cracklings), ground

3 cloves garlic
1 cup stock or broth
1 tablespoon olive oil
Vegetable oil, for frying

Peel the plantains and cut into 1-inch diagonal pieces.

Heat the oil in a large skillet. Place the plantains in the oil and fry on both sides, approximately 3½ minutes per side. Remove the plantains from the pan and flatten the plantains by placing a plate over the fried plantains and pressing down. Return them to the hot oil and fry, 1 minute on each side. Place on paper towel.

Mince garlic. While the plantains are still hot, use a mortar and pestle to mash them with the minced garlic, olive oil, and stock or broth. Add the chicharrones, salt and pepper to taste.

Note that you can also use a food processor: add the plantains to the food processor with bacon, garlic and some salt and pepper. You may have to work in batches. Process to consistency of mashed potatoes. Be careful not over-process!

Note: For this recipe, packaged pork rinds can also be used, as can any kind of bacon.

Simple Sweet and Sour Pork

This is a recipe to use with leftover pork. Adjust to taste and the amount of meat you're using.

½ cup brown sugar
1 tablespoon cornstarch
⅓ cup red wine vinegar
1 tablespoon soy sauce
20 ounce can of pineapple
  chunks in unsweetened
  juice (not syrup)

Maraschino cherries, to taste
Cooked pork, to taste
Celery, to taste

Mix the brown sugar and corn starch. Add the vinegar, soy sauce, and most of the pineapple juice. Heat to a boil, while stirring constantly. Reduce heat and add pre-cooked pork, pineapple chunks, cherries, and celery. Cook until celery begins to soften.

Note: Chicken may also be substituted.

Babi Kecap: Pork Braised with Garlic and Onions

Serves 6

2 pounds pork
5 large white onions
5 cloves garlic
4 teaspoons soy sauce

Ginger, fresh or powdered,
    to taste
Black pepper, to taste
Chili pepper (optional)

Cut the onions and cloves of garlic. Cut the meat into 1-inch cubes. Add the onions, garlic and ginger into a pan and cook on high. Add the pork and mix with the soy sauce and black pepper (and with chili pepper, if desired).

After about 3–4 minutes, reduce the heat to medium and continue to cook, covered, for one hour. Serve with rice of your choice.

Wine Glazed Pork Chops

4 thick cut bone-in pork rib chops
½ cup Chop Rub (see below)

Cider vinegar, as needed
1 cup red wine
¼ cup butter, divided

Brush pork chops generously with cider vinegar and season with Rub, patting gently.

Bring wine to a boil over high heat in a large saucepan and ignite. Cook until reduced by half. Add butter 1 tablespoon at a time while whisking, until all butter has melted.

Grill pork chops over high heat for 2 minutes, brushing with wine mixture, then twist them 90 degrees and cook for another 2 minutes, brushing. Flip and repeat 1 more time. Remove and let rest 5 minutes. Serve with remaining wine mixture.

Chop Rub

May use on chicken beef or pork

1 cup smoked paprika
½ cup salt
1 cup dark brown sugar
¾ cup dehydrated lemon peel
½ cup black pepper
¼ cup dried rosemary

⅓ cup granulated garlic
½ cup chipotle powder
2 teaspoons hickory smoke powder (or 2 drops liquid smoke)

Combine all ingredients and store in a cool, dark place in an airtight container.

Apple Stuffing

A great complimentary dish with pork.

1½ cups cornbread, crumbled finely
½ cup dried apples
½ cup apple jelly
¼ cup walnuts, roughly chopped
2 tablespoons sugar
4 egg yolks, beaten

4 egg whites, beaten to stiff peaks with ⅛ teaspoon cream of tartar
¼ cup shredded Fontina cheese
Kosher salt and freshly ground black pepper, to taste

Combine all ingredients except cheese and egg whites. Gently fold in egg whites, in batches.

Pour into a roasting pan. Top with shredded cheese and bake at 325°F for half an hour or until browned and bubbly on top. Remove and let rest for 7 minutes.

Asian Barbecued Pork: Char Sui

A very simple barbecue seasoning recipe for pork.

4 pounds of pork
  (sam-chang-bak)
5 tablespoons of char-siu sauce
5 tablespoons of sugar

Freshly ground black pepper,
  to taste
A few drops of tenderizer

Sprinkle the tenderizer on the pork, and then mix the rest of the ingredients together with the pork. Grill as desired.

Chipotle Maple Glazed Pork Tenderloin
(For the Grill)

1 pork tenderloin, trimmed of silver skin

4 chipotle chiles in adobo sauce, finely minced (easy to find at most grocery stores)

½ cup real maple syrup

¼ cup freshly squeezed lime juice

1 tablespoon finely grated lime zest

1 teaspoon garlic powder

1½ teaspoons salt

1½ teaspoons freshly ground black pepper

2 tablespoons honey

Combine lime, garlic, salt, honey, and pepper in a gallon size resealable bag. Add pork and refrigerate 24 hours, then drain pork and set aside.

Heat 4½ pounds of (preferably) natural chunk charcoal in a large chimney starter. Once coals have heated, disperse evenly around the bottom of the grill. Reapply grill grate and swab grate with an oil-soaked towel to help prevent pork from sticking to the grill.

Add the pork to hottest part of grill and cook 1½ minutes. Turn 90 degrees and cook another 1½ minutes. Repeat 12 times, or until internal temperature reaches 140°F for medium rare.

While the pork cooks, combine chipotle and maple syrup. Set aside.

Make a foil pouch and place cooked pork inside. Pour maple mixture over and tightly seal. Let rest 15 minutes and serve warm.

Steamed Buns with BBQ Pork

Steamed Buns
1½ cups warm water (95°F)
3 teaspoons active dry yeast
3 tablespoons granu-
lated white sugar

4 cups white bread flour
2 tablespoons cooking oil
1 teaspoon baking powder
½ teaspoon salt

Bun Filling
8 ounces shredded cooked
pork
½ cup barbecue sauce of your
choice
1 tablespoon soy sauce
1 tablespoon cooking oil

1 tablespoon granulated white
sugar
1 tablespoon Teriyaki sauce
1 teaspoon garlic chili sauce
¼ teaspoon five spice powder
¼ teaspoon toasted sesame oil

Filling
Place all ingredients in food processor. Process until ingredients
are blended and chopped fine.

Use 1 tablespoon of mixture per bun.

Steamed Buns
Stir yeast and 1 tablespoon of sugar into 1½ cups of warm
water (95°F). Let stand for 15 minutes. Mix yeast water, 2 more
tablespoons of sugar, flour, cooking oil, baking powder and salt
in mixing bowl. Stir until dough forms. Knead dough for 10
minutes. Allow dough to rise in a warm place for 1½ hours.

Punch dough down and divide into 12 pieces. Roll each piece
of dough into a ball, and then flatten with rolling pin into a 6-inch
(15cm) circle. Place 1 tablespoon of meat filling in center of each
flattened piece of dough. Lift the edges of the flattened dough up
around the filling and bring edges together. Pinch edges closed to
form a filled bun.

Place buns on parchment paper. Allow to rise in a warm place for 1 hour, or until doubled in size. Steam buns in steamer over boiling water for 20 minutes. Remove from steamer. When cool enough to handle, remove parchment paper from bottom. Serve warm.

Memphis-Style Ribs

1 slab pork spareribs
6 tablespoons Rib Rub
  (see below)
1 cup tomato paste
½ cup molasses

¼ cup apple cider vinegar
2 tablespoons Worcestershire
  sauce
2 tablespoons soy sauce
Hickory chunks

Season spareribs with rib rub and refrigerate for at least 1 hour. Place chunks in smoker or charcoal grill set for indirect heat at 250°F.

Combine liquid ingredients and bring to a boil over high heat. Reduce liquid by one-third. Place spareribs into smoker or grill for 4-5 hours, basting occasionally with sauce. Brush remaining sauce on top of ribs and place on a medium high grill until browned. The internal temperature of the ribs before removing from the grill needs to be at least 145°F. Let rest for 10 minutes, covered, and carve.

Rib Rub

2 cups smoked paprika
¼ cup salt
1 cup brown sugar (dark or
  light)
3 tablespoons cayenne pepper
  (or to taste)

¼ cup dry mustard
¼ cup garlic powder
¼ cup dried rosemary
1 tablespoon cinnamon

Combine all ingredients in a container with a shaker lid. Store leftover rub in cool, dry place.

Asian Barbecue Ribs

2 slabs St. Louis style/cut
  pork spareribs (or personal
  preference)
6 tablespoons Barbecue Rub
  (see below)
¼ cup soy sauce

¼ cup hoisin sauce
½ cup sake
3 tablespoons honey
1 tablespoon mirin
Dry hickory chunks

Peel off the thin membrane on each rib. Season ribs with Barbecue Rub and refrigerate for at least 1 hour.

Place chunks in the firebox of a 250°F smoker. Combine remaining ingredients in a large sauce pan over high heat and bring to a boil. Cook, stirring occasionally, for 5 minutes or until slightly thickened. Set aside.

Place ribs in smoker and cook for 5–6 hours, brushing with soy mixture often. Remove from smoker. Brush meat sides generously with soy mixture and place under a preheated broiler until deeply browned. Watch carefully, or it will burn! Slice into single rib portions and serve.

Barbecue Rub

1 cup brown sugar
¼ cup salt
¼ cup garlic powder
1 cup paprika
¼ cup chipotle powder

3 tablespoons cayenne pepper
3 tablespoons dried rosemary
¼ cup chili powder
¼ cup dehydrated lemon peel

Combine all ingredients in an airtight container. Store any leftover rub in a cool, dry place.

North Carolina-Style BBQ Ribs

1 slab pork spareribs
6 tablespoons Rib Rub
  (see page 124)
1 cup apple cider vinegar

⅓ cup honey
2 tablespoons Worcestershire
  sauce

Season ribs with Rib Rub and refrigerate for at least 1 hour. Combine remaining ingredients and place in a spray bottle.

Place ribs in a 250°F smoker or a charcoal grill set for indirect heat at 250°F. Cook for 5–6 hours, spraying often with vinegar mixture. Spray heavily with vinegar mixture and place on a high grill, turning and spraying constantly until well browned. Remove and let rest 15 minutes; serve.

Note: There will be extra vinegar mixture. Don't throw it away! You can use it for almost anything, including beef briskets.

RESOURCES

Online

Merck Vet Manual
www.merckvetmanual.com

Formulating Farm-Specific Swine Diets
www.extension.umn.edu/agriculture/swine/formulating-farm-specific-swine-diets/

The Pig Site Quick Disease Guide
www.thepigsite.com/diseaseinfo/

The New Century Homesteader
www.newcenturyhomesteader.blogspot.com

Books and Articles

Morton Salt Company. (1975). *A Complete Guide to Home Meat Curing*. (1975). Chicago, Ill.

Belanger, J. (1977). *Raising the Homestead Hog*. Emmaus, Pa.: Rodale Press.

Loon, D. (1978). *Small-Scale Pig Raising*. Charlotte, Vt.: Garden Way Pub.

Klober, K. (2009). *Storey's Guide to Raising Pigs: Care, Facilities, Management, Breeds* (3rd ed.). North Adams, MA: Storey Pub.

Raising Pigs of Your Own. (May/June 1980). *Mother Earth News*.

Raising Pigs for Meat, (March/April 1970). *Mother Earth News*.

How to Raise Pigs on Pasture. (April 2014). Smith Meadows, www.smithmeadows.com.

Sheely, W. (1932). *Butchering and curing pork*. Gainesville, Fla.: Cooperative extension work in agriculture and home economics.

Organizations

National Pork Board
www.pork.org

Magazines

Countryside Small Stock Journal
www.countrysidemag.com
A series of magazines, which contain a variety of good articles on raising and keeping pigs.

Mother Earth News
www.motherearthnews.com

Also in the *Backyard Farming* Series…

Backyard Farming: Canning & Preserving

Backyard Farming: Growing Garlic

Backyard Farming: Growing Vegetables & Herbs

Backyard Farming: Home Harvesting

Backyard Farming: Keeping Honey Bees

Backyard Farming: Raising Cattle for Dairy and Beef

Backyard Farming: Raising Chickens

Backyard Farming: Raising Goats

Backyard Farming: Homesteading

NOTES

NOTES

NOTES

_____
_____
_____
_____
_____
_____
_____
_____
_____
_____
_____
_____
_____
_____
_____
_____
_____
_____
_____
_____
_____
_____
_____
_____

NOTES

NOTES